T0259787

Hydroponics for the
HOME GROWER

Hydroponics for the
HOME GROWER

Howard M. Resh, PhD

CuisinArt Golf Resort & Spa
Anguilla, British West Indies

CRC Press
Taylor & Francis Group
Boca Raton London New York

CRC Press is an imprint of the
Taylor & Francis Group, an **informa** business

CRC Press
Taylor & Francis Group
6000 Broken Sound Parkway NW, Suite 300
Boca Raton, FL 33487-2742

Printed on acid-free paper
Version Date: 20150623

International Standard Book Number-13: 978-1-4822-3925-6 (Paperback)

Library of Congress Cataloging-in-Publication Data

Resh, Howard M., author.
   Hydroponics for the home grower / Howard M. Resh.
      pages cm
   Includes bibliographical references and index.
   ISBN 978-1-4822-3925-6 (pbk. : alk. paper) 1. Hydroponics. I. Title.

SB126.5.R494 2015
631.5'85--dc23
                                             2014042946

Visit the Taylor & Francis Web site at
http://www.taylorandfrancis.com

and the CRC Press Web site at
http://www.crcpress.com

Contents

SECTION I   History and Background of Hydroponics

SECTION II   Understanding Hydroponics and How Plants Grow

SECTION III   Nutrients Essential to Plants and Their Sources

SECTION IV   Hydroponic Systems

SECTION V   Year-Round Growing in Greenhouses

SECTION VI   Vegetable Crops and Their Cultural Techniques

SECTION VII   Sprouts and Microgreens

List of Figures

List of Tables

Acknowledgments

This book gathers information from almost 40 years of my experience with hydroponics. During the early 1970s people were attempting to design small indoor hydroponic units, but with little knowledge of the general public of what hydroponic growing was all about, most of these business ventures into marketing of small indoor units were not successful. It took almost a decade for consumers to become more aware of hydroponics and its potential for indoor growing. Companies established at this later period in marketing such small indoor units did become very successful, and, in fact, many of them still exist today. Now they manufacture and sell all components from lights, carbon dioxide generators, nutrients, and so on, in addition to the hydroponic growing units themselves. Some of these companies now have world-wide distributors of their products.

In this book, many of these products are presented as all components of growing are key to successful indoor growing. Some of the companies that manufacture and/or sell these units and accessories mentioned in this book are listed in the Appendix. I wish to thank the following companies for the use of their photos in this book: AeroGrow International, American Hydroponics, AutoPot Global, Ltd., Bluelab Corporation Ltd., Botanicare, CO2Boost LLC., General Hydroponics, Green Air Products, Inc., Hydrofarm Horticultural Products, Hydrofogger, LumiGrow, Milwaukee Instruments, Myron L Company, Sunlight Supply, Inc.

A very special thanks to Mr. Leandro Rizzuto, President of CuisinArt Golf Resort and Spa, Anguilla, British West Indies, for permitting me to include many photos that I have taken while working at the CuisinArt Hydroponic Farm.

In no way is the use of trade names intended to imply approval of any particular source or brand name over other similar ones not mentioned in this book.

Introduction

Most of us growing vegetables in our backyard gardens face lots of challenges with the soil structure, fertility, watering, pests, and diseases. You may think that all you have to do is to sow some seeds in the soil and they will germinate and grow into productive plants. This, however, is wishful thinking, unless you know the proper techniques for successful soil growing. So, if you are faced with these challenges, is there another way to give you better control of these limiting factors to your production? The answer is hydroponic culture.

Basically, growing plants in soil or hydroponically requires similar needs from their environment for good yields. However, with hydroponics you have control over many of the limiting factors that plants encounter in soil. One step further is to grow them in a greenhouse hydroponically. In this way, you can also control some of the outside limiting factors, such as light and temperature, and exclude to a large degree pest and disease problems with your plants. In addition, you can grow year-round in a greenhouse, producing high-quality plants even during winter months. Experience production during the winter and escape from the darkness and stress of winter doldrums during these short winter days.

ABOUT THIS BOOK

In this book, focus is on the production of vegetable crops year-round within your home or in a greenhouse. By following the procedures presented, you can grow successfully with hydroponic culture. While most crops grown in soil can be grown hydroponically, the emphasis, due to economic viability, is on tomatoes, peppers, cucumbers, eggplants, lettuce, arugula, bok choy, and various herbs. I present background information on how hydroponics evolved, plant needs in terms of nutrients, water, plant growth, and show you how you can provide these basic needs to your plants. There is an explanation of nutrient solution makeup to show you how to provide the plants with their essential elements for growth. Nutrient solution formulations and their preparation are basic to successful hydroponics. However, they can be purchased from hydroponic shops and online if you wish to avoid making them up yourself.

After that, many hydroponic systems are described that you may construct yourself or purchase. These systems, with their substrates, are taking the place of the soil. They can be automated to reduce your constant caring as occurs with soil in providing fertilizers and building soil structure and fertility through composting and watering.

I take you one step further to enjoy gardening year-round with the use of a backyard greenhouse. The construction of backyard greenhouses is described along with the components needed to control the climatic conditions within the greenhouse that are favorable to plant growth. Hydroponic systems for these backyard greenhouses are the next step in assuring successful growing for the whole year.

Finally, numerous vegetable crops recommended for hydroponics are discussed in detail, including such topics as seeding, transplanting, training of the plants, and pest and disease control. Varieties I have found in the past that grow best under greenhouse hydroponics are given, along with some simple indoor systems to grow sprouts and microgreens.

TERMINOLOGY USED IN THIS BOOK

Temperatures are given in the Fahrenheit (F) and Celsius (C) scales as many countries now use Celsius. I place the Celsius temperatures in brackets after the Fahrenheit ones. *Italics* are used for plant, insect, and disease Latin names. Scientific weights and measurements will be defined as they are introduced. Other shortened forms will be placed in brackets immediately after the word(s) when first introduced. For example, electrical conductivity, basic or acid measurement, nutrient film technique (NFT), and so on.

WHY HYDROPONICS IS FOR YOU

The way I look at it, if you must spend considerable time caring for your plants in soil, why not eliminate some of the variables that are restricting good yields in your backyard by going to hydroponic culture. Hydroponics is quite logical and only requires step-by-step procedures. Doing so will greatly increase your growing success. It also is less "back breaking" work than occurs with your soil garden in weeding, hoeing, mulching, and adding soil supplements, such as steer manure, fertilizers, and so on. And you can avoid many soil-borne pests and diseases, so less spraying of pesticides is necessary. Hydroponic systems can be constructed at waist height to save on bending over to look after and harvest plants, especially low-profile ones like lettuce, bok choy, cabbages, spinach, strawberries, and herbs. You can even grow root crops like carrots, onions, and green onions in some forms of hydroponics, such as a peat-lite mixture or coco-coir substrate. Bush beans also grow well in raised beds of these media. All your efforts will be well rewarded in higher yields of your crops.

DO NOT BE FOOLED BY "ORGANIC" PLANTS

Many people believe that hydroponically-grown plants are not organic. Of course, that is not true. In fact, all plants are organic. They all require elements essential for their growth (essential elements), including carbon from carbon dioxide and oxygen and hydrogen from the air and water. There are no "organic" elements required that they only receive in soil. In fact, organic compounds (those containing carbon) must be broken down into their elemental constituents to be absorbed by plants. These organic compounds are in the form of decaying plant and animal material that through microbial decomposition release their elements in atomic (ionic) states into the soil water to form the soil solution. The plant roots are in contact with the soil solution and take up the essential elements by expending energy to transport them across their root membranes. The soil is also composed of inorganic compounds such as sand, rocks, and so on, that must be weathered to break down into their elements. Once again they are released into the soil water, resulting in the soil solution.

In hydroponics, we dissolve essential element-bearing compounds in water to form the nutrient solution. The nutrient solution serves the same function as the soil solution in providing the essential elements (13 of them) to be available to plant roots. The growth of the plants is the same whether in soil or a soilless medium. Plants through photosynthesis and respiration using carbon, hydrogen, and oxygen from the air and water manufacture their building blocks for growth.

"Organic" growing is really a misnomer as it really applies to the non-use of synthetic pesticides. These very same natural pesticides and beneficial insects are used in hydroponic growing, but they can be more efficiently applied and controlled under hydroponic culture than in soil culture. One step further is to do the growing in a greenhouse, where the pests and diseases can be excluded or restricted to some extent, and the release of beneficial insects in the closed environment of the greenhouse keeps them within the greenhouse, where they can multiply while controlling the pests.

ORGANIZATION OF THIS BOOK

This book is divided into sections. Each section covers a number of chapters related to a theme. The following is an outline of the parts.

SECTION I: HISTORY AND BACKGROUND OF HYDROPONICS

Hydroponics, while not termed that until the 1940s, was practiced by ancient cultures. As scientists later looked for the reasons behind plant growth and their needs for development, they used various forms of "nutriculture" (growing plants in substrates other than soil) to discover these factors. Knowing some of this background will help you to understand that hydroponics is not something that just developed overnight. It took many centuries of study to finally apply it to commercial growing. Where this culture is now and where it is heading in the future will provide you with insight to its many applications. This leads to its popularity and benefits that will convince you of its advantages for your growing hydroponically even on a small scale.

SECTION II: UNDERSTANDING HYDROPONICS AND HOW PLANTS GROW

This section will demonstrate that hydroponics is not just all chemistry but still regular gardening with a twist of providing plants with all of their components at more optimal levels in order to improve yields. Understanding some of the nutritional and environmental demands of plants will enable you to recognize and provide these factors for them. Next, then, is how to do this by proper watering (irrigation), nutrient application and levels, and management of the environment to make your plants happy and productive.

SECTION III: NUTRIENTS ESSENTIAL TO PLANTS AND THEIR SOURCES

While this is emphasizing hydroponics, it also applies to soil gardening. I compare hydroponics with soil growing in terms of where and how plants get their nutrients. Sources of the nutrients will assist you in finding the purist, high-quality compounds that will make these nutrients readily available to your plants. With hydroponics, you have control of providing your plants with optimal levels of nutrients (essential elements) by using specific formulations optimal to those plants you wish to grow. Emphasis is always on vegetable crops, as these are the ones that you want to maximize production. Finally, when you understand nutrition of the plants, you can observe any disorders that may occur as a result of nutrients in deficiency or excess that cause specific symptoms with the plants. When the plants are under such stress, yields will fall. Upon recognizing these shortfalls in nutrition, you are shown how to cure them.

SECTION IV: HYDROPONIC SYSTEMS

This is a real fun section as it will show you many hydroponic systems that you can build yourself or purchase. You will learn the substrates or media (other than soil) that can be used in hydroponic growing, sometimes called "soilless" culture. The characteristics of such substrates will help you to determine which form you may wish as your growing medium. Their sources and which plants thrive best with certain media help you to decide on what you should use. We start with small indoor units for your home and progress to larger ones. Do-it-yourself (DIY) designs and construction of these hydroponic units help you to decide where to start. From there I explain how to start your own plants from seed and provide you with the components you need, such as seeds, trays, substrate, and so on, and their sources. How to choose the hydroponic system for the specific crops you wish to grow is important to enhance your success as some plants prefer certain growing systems over others. Low-profile plants like lettuce and herbs are better adapted to some hydroponic systems than vine crops that need more rooting space, doing better in containerized systems. Sources of these components and the types of each you should look for will make your search easier when constructing your system or purchasing one.

SECTION V: YEAR-ROUND GROWING IN GREENHOUSES

This is a very rewarding hobby and an excellent way of getting away quickly from the winter doldrums. You can escape to a summer paradise of plants in your greenhouse during the height of winter. As pointed out earlier, to grow plants to their maximum benefit, you must provide the best possible environment for them. Those factors include temperature, light, relative humidity, irrigation, and so on that can be accomplished within a greenhouse. There are many benefits of a greenhouse to the homeowner from clean, healthy, nutritious vegetables to education for children to a psychological uplift. Presented are designs for different types of greenhouses and sizes to fit your personal taste and budget. Construction for DIY projects and commercially available prefabricated structures with all the components are presented. Sources and approximate costs of structures and environmental components, such as heaters, fans, lights, and so on, are all part of the knowledge of greenhouses. Designs, sizes, types of hydroponic systems, DIY construction, and sources of components and supplies will assist you in growing the crops you wish.

SECTION VI: VEGETABLE CROPS AND THEIR CULTURAL TECHNIQUES

Information is given to guide you on the crops most commonly grown hydroponically and how to select them from the many varieties. From there step-by-step procedures of seeding and transplanting clearly guide you through these processes. You are shown which growing supplies are best for specific cultures of individual crops. After this are detailed descriptions for training your crops, such as suckering tomatoes, peppers, cucumbers, and eggplants, supporting these plants vertically, pollination, lowering and leaning the plants, and other cultural practices specific to each crop. The next chapter tells you when to plant, how long from seed to first harvest, and when to change the crops in terms of cropping cycles to best suit your growing conditions. The management of recognizing and controlling pests and diseases in the following chapter is crucial to your success, whether in soil or soilless growing. Treating each crop for its most common pests and diseases and the specific biological or pesticide controls will keep your plants healthy and productive.

SECTION VII: SPROUTS AND MICROGREENS

The emphasis of this section is on the simplified growing of microgreens and why to grow them instead of sprouts. This is a very safe crop that can be grown on your kitchen countertop. The descriptions of straightforward procedures to grow microgreens, with sources of supplies, nutrients, trays, seeds, lights, and so on, makes this a winner in short-term growing within 5–12 days from seeding. Besides, it is a great science project in hydroponics for school classes.

FINAL ADVICE ON HOW TO GET STARTED

This section presents a simplified summary of events to follow for establishing your hydroponic gardening.

APPENDIX: SOURCES OF SUPPLIES AND INFORMATION

Guidelines are presented as tables for sources of seeds and other supplies in your area. Websites are listed on hydroponics, vegetable culture, pest and disease management, backyard greenhouses, greenhouse components, university extension agents, and so on to easily seek further information. A reference is provided of books, articles, and conferences on growing hydroponically as well as caring for your crops.

Author

Howard M. Resh is a recognized authority worldwide on hydroponics. His website (www.howardresh.com) presents information on hydroponic cultures of various vegetable crops. In addition, he has written five books on hydroponic culture for both commercial and hobby growers.

Upon graduation with his doctorate in horticulture in 1975, he became urban horticulturist for the faculty of plant science at the University of British Columbia. He held that position for three years before the call of commercial hydroponics took him to projects in Venezuela, Taiwan, Saudi Arabia, the United States, and the British West Indies, in 1999, where he is today.

While in the position of urban horticulturist, Resh taught courses in horticulture, hydroponics, plant propagation, and greenhouse design, and production. During this period and later while general manager for a large plant nursery, he continued researching and consulting for a commercial hydroponic farm growing lettuce, watercress, and other vegetables in Venezuela. Resh became project manager for the Venezuelan farm to develop hydroponic cultures of lettuce, watercress, peppers, tomatoes, and European cucumbers using a special medium of rice hulls and coco coir from local sources. He also designed and constructed a Mung bean and alfalfa sprout facility to introduce sprouts into the local market.

In the late 1980s, Resh worked with a company in Florida in the growing of lettuce in a floating raft culture system.

From 1990 to 1999, Resh worked as the technical director and project manager for hydroponic projects in the growing of watercress and herbs in California. He designed and constructed several 3-acre outdoor hydroponic watercress facilities using a unique NFT system. These overcame production losses due to drought conditions in the area.

From there in mid-1999, Resh became the hydroponic greenhouse farm manager for the first hydroponic farm associated with a high-end resort, CuisinArt Golf Resort and Spa in Anguilla, British West Indies, in the Northeastern Caribbean. The hydroponic farm is unique in being the only one in the world owned by a resort growing its own fresh salad crops and herbs exclusively for itself. This farm has become a key component of the resort in attracting guests to experience home-grown vegetables including tomatoes, cucumbers, peppers, eggplants, lettuce, bok choy, and herbs. The resort, together with its hydroponic farm, has gained world-wide recognition as one of the leading hotels of the world.

Resh continues to do consulting on many unique hydroponic greenhouse operations, such as Lufa Farms in Montreal, Canada. There he established the growing techniques and hydroponic systems for a rooftop hydroponic greenhouse in downtown Montreal. All vegetables are marketed through a community-supported agriculture program.

Section I

History and Background of Hydroponics

1 How Hydroponics Started

Its Present and Future Applications

During ancient times, people faced many challenges while gardening. Being the curious animal man is, he wanted to find out what made plants grow. Soil was a mysterious material that somehow provided the right conditions for plants to grow from seed into plants that produced edible parts. Often plagues and pestilence reduced or even destroyed the yields of plants that societies heavily depended on for their well-being. When crops failed, societies suffered famine and death. Such crop failures led to wars between neighboring communities and even the death of entire civilizations and cultures. I heard the phrase "No Agriculture, No Culture," recently on a TV historical documentary. This statement clearly points out the fact that cultures and civilizations are dependent on crops for their survival. If man knew more of the causes of these crop failures, he could try to prevent them. This became the basis of agriculture—to find out the reasons for plants to thrive so that man could cultivate plants under favorable conditions, which would lead to abundant production.

Ancient civilizations became aware that water was essential for any agricultural practices, so populations gathered in areas that had an abundant source of water that could be used for growing plants. Usually, by streams, rivers, lakes, or springs that had fresh water, civilizations developed where they could practice agriculture. Fertile soil existed in valleys of rivers and near lakes. Such soil supported productive crops and human centers. When groups of inhabitants experienced harsh environments that restricted their agricultural crops, they needed to examine what factors reduced yields and what could be done to improve them.

In the early times, man became aware of growing plants in specific environments and tried new methods of cultivation both for ascetics and food. Egyptian hieroglyphic records of several hundred years BC describe growing plants in water. Theophrastus during 372–287 BC experimented with plant nutrition. A form of hydroponics was established with the hanging gardens of Babylon, the floating gardens of the Aztecs of Mexico, and the Chinese. However, these were not called "hydroponic" culture even though they were a form of it.

Further experiments with a scientific approach to discover plant constituents were carried out by numerous scientists during the 17th century and later. They were able to discover that water, soil, and air provided elements such as carbon, hydrogen, and oxygen that were constituents of plant matter. Researchers later continued to demonstrate that the minerals that plants contained came from the soil via the soil water.

3

This enabled scientists to later grow plants in water alone without soil provided that these minerals were added to the water.

This became "nutriculture," where plant roots were immersed in a water solution containing salts of their essential elements.

From 1925 to 1935, laboratory-scale nutriculture was expanded to commercial-scale production of crops. However, it was not until the 1930s and 1940s that the application of nutriculture was applied on a commercial scale by Dr. W. F. Gericke of the University of California and termed "hydroponics." The word "hydroponics" was derived from two Greek words *hydro* ("water") and *ponos* ("labor")—"water working."

In the 1940s, with the war in the Pacific, Gericke applied hydroponics to commercial production in the nonarable islands where troops were stationed. After the war, hydroponic culture was adopted by the greenhouse industry to resolve problems with soil-borne diseases and pests as well as structural and nutritional challenges faced by year-round growing in greenhouses (Figure 1.1). Now, almost all crops grown in greenhouses, including vegetables and ornamentals, use some form of hydroponics. It may also be termed "soilless culture" when using an inert medium other than soil to which a nutrient solution is added.

Hydroponic greenhouse growing is now worldwide. Some of the largest vegetable greenhouse production regions include Holland, Spain, England, Canada, United States, Mexico, Turkey, China, Australia, and Middle Eastern countries. Holland has more than 25,000 acres of greenhouse production, which includes ornamentals and flowers. Canada has about 2800 acres of greenhouse hydroponic

Soil–small plant
big root system

Hydroponic–smaller root
system more healthy and
more productive

FIGURE 1.1 Comparison of plants growing in soil versus in a soilless system. (Drawing courtesy of George Barile, Accurate Art, Inc., Holbrook, New York.)

vegetable production and the United States 1500 acres. China is rapidly expanding its greenhouse production with presently approximately 3100 acres. Other areas of expansion include Turkey, Mexico, Morocco, and Australia.

Hobby hydroponic culture started in the 1940s and 1950s with gravel and water culture systems. These were mainly "Do-It-Yourself" projects. In the 1970s, some of the first commercially available hobby hydroponic units entered the marketplace as automated systems to simplify hydroponics for households. The "City Green" hydroponicum was one of the first such units constructed of molded plastic with an upper growing tray and a nutrient reservoir below. The substrate was volcanic cinder rock or expanded clay irrigated by a small perforated plastic tube on the top of the medium. A tube from a fish aquarium pump outside was connected to the irrigation tube in the nutrient tank. The air pump tube was connected to the larger diameter irrigation tube. The space between the walls of the tubes at their connection permitted water to move up by the force of the air entering the irrigation tube as shown in Figure 1.2.

Presently, with the increased interest in home hydroponics, a vast number of designs and types of systems are marketed for all types of crops (Figures 1.3 and 1.4). They are available online and/or in hydroponic outlets in most countries. Specific types of units and their application to most suitable crops is discussed later in Chapters 12, 13, and 15. In the future, with increased awareness of food quality and safety, I am sure the general population will adapt hydroponic growing in their households, especially for herbs and salad crops.

Commercial hydroponics in the future will become associated with tourist resorts and spas as they are emphasizing wellness programs for their guests. Industries with waste heat and geothermal sites will couple with hydroponic greenhouses to produce vegetables more economically by using cheaper sources of energy for heating. Increased efficient light sources, such as light emitting diode (LED) lights, are

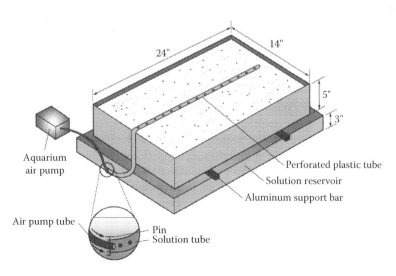

FIGURE 1.2 Components of an indoor unit. (Drawing courtesy of George Barile, Accurate Art, Inc., Holbrook, New York.)

FIGURE 1.3   "Aerogarden" kitchen unit. (Courtesy of AeroGrow, Boulder, Colorado.)

FIGURE 1.4   Kitchen countertop garden ebb-and-flow unit. (Courtesy of American Hydroponics, Arcata, California.)

rapidly entering the greenhouse industry in northerly latitudes where light is limited during winter months. This trend will continue with new sources of lighting.

Hydroponic greenhouse operations are now being established on roof tops of buildings in the centers of cosmopolitan cities (Figure 1.5). Such operations now exist in Montreal, New York, and are presently expanding to Vancouver and New Jersey. Another approach is to locate greenhouses in parking lots adjacent to supermarkets.

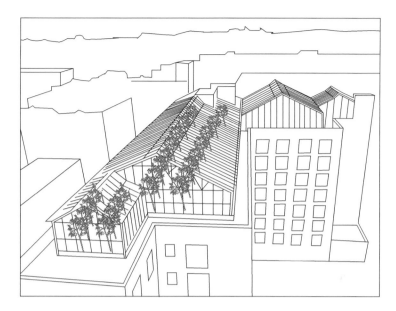

FIGURE 1.5 Greenhouse hydroponic farm on a rooftop in the city. (Drawing courtesy of George Barile, Accurate Art, Inc., Holbrook, New York.)

These applications of hydroponics provide clients with fresh produce, free of toxic pesticides, and fully vine-ripened fruits such as tomatoes, peppers, and eggplants. The other factor is the saving of fossil fuels in long transportation for distant markets. The product is grown onsite at the retail outlet or with community-supported agriculture marketing where fresh vegetables are taken to nearby drop-off points for "subscribed" consumers to pick up. The concept is to have households sign up as a member and then pay a monthly fee for their vegetables that are in returnable baskets that are either picked up at the drop-off site or can be acquired at the greenhouse operation itself on specific days.

Going one step further into the future, I expect that high-rise vertical buildings will be constructed in city centers to grow vegetables. They could also be part of a condominium complex where some floors or a wing of the building would be modified to grow plants with highly efficient hydroponic systems, such as rotating, vertical structures. This technology of rotating, vertical hydroponic systems already exists. However, the success of these high-rise greenhouses is dependent on a very efficient source of supplementary lighting, so I believe that it could happen within a decade or so. Solar cells on the rooftop of the building or in a nearby parking space could provide the electrical needs for the lights.

Hydroponics opens up potential for growing crops under all environmental conditions including in your home and/or backyard. It is the wave of future growing for you, so be part of it!

2 The Popularity and Benefits of Hydroponic Gardening

The versatility of hydroponics makes it popular throughout the world from simple household, backyard applications through commercial greenhouses to isolated locations in the Antarctic and under zero gravity on the space station (Figure 2.1). Someday, as we probe space exploration and set foot on other planets, we will cultivate food crops under special greenhouse structures with hydroponic culture. This is already exemplified in scientific papers and even in science-fiction movies. In the future of space exploration, it will become a reality.

At the other end of the spectrum, hydroponics is a feasible culture for low-income societies. Very simplified hydroponic systems using waste materials and basic supplies of nutrients are now known as "popular hydroponics" in Latin America (Figure 2.2). In desert regions of Peru, people living under harsh conditions of existence have turned to satisfying some of their nutritional needs through hydroponic culture. Often, for example, this is a result of some assistance by local universities, such as Universidad de La Molina in Lima, Peru, and the Food and Agriculture Organization of the United Nations. These institutions provide classes and subsidized supplies for low-income rural people to initiate and carry out hydroponic culture of their basic vegetables for a more healthy diet. Often these societies come together to produce larger facilities on roof tops of schools and community centers and operate as a cooperative in exchanging produce (Figure 2.3). It also becomes an educational facility for school students. I have personally visited numerous sites of this nature in Peru and seen how people who started just to supplement their diets have expanded to become commercial operations and now make a living by growing vegetables for markets in the large cities such as Lima. This same process has happened in other countries as Colombia, Venezuela, Bolivia, Uruguay, and Brazil.

Hydroponics has become popular in all societies as people learn of the many benefits to grow crops free of weeds (Figure 2.4), in control of pests and diseases, and obtain high yields of highly nutritious and safe vegetable crops. I am not trying to mislead you by suggesting that hydroponics is the answer to future food shortages. The point is that you can grow some of the most nutritional vegetables hydroponically such as tomatoes, peppers, lettuce, herbs, and so on that provide healthy products with the least environmental footprint. That is a win–win situation for you and the environment. Do not be left out, this is the way of the future!

The benefits of hydroponics over soil culture are great. Increased demand for food production has to focus on more efficient methods of water usage,

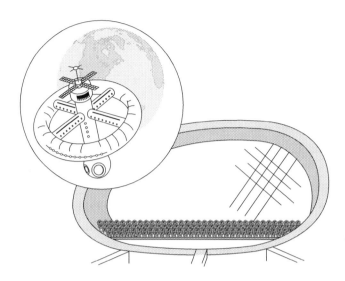

FIGURE 2.1   A hydroponic garden on a space station. (Drawing courtesy of George Barile, Accurate Art, Inc., Holbrook, New York.)

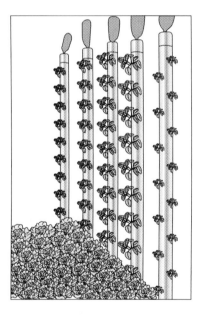

FIGURE 2.2   Simple (popular hydroponic) gardens for poor community backyards. (Drawing courtesy of George Barile, Accurate Art, Inc., Holbrook, New York.)

less dependency on toxic pesticides, higher yields, and superior quality of products in both flavor and nutrition. These factors contribute to less demand on our environment making hydroponic culture very "green." In hydroponic greenhouse operations, the emphasis is on "sustainable" agriculture with a minimum environmental foot print.

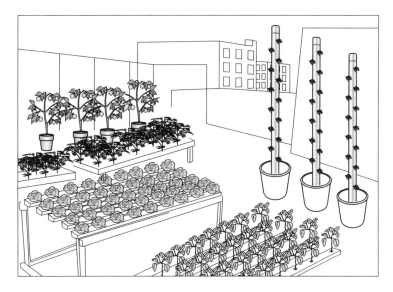

FIGURE 2.3 Simple hydroponic rooftop gardens. (Drawing courtesy of George Barile, Accurate Art, Inc., Holbrook, New York.)

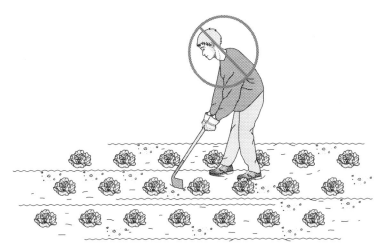

FIGURE 2.4 Avoid bending over to hoe and weed in a normal soil garden. (Drawing courtesy of George Barile, Accurate Art, Inc., Holbrook, New York.)

Some of the obvious advantages of hydroponics over soil culture include the following:

1. Ease and efficiency of sterilization of the medium between crop turnarounds.
2. Nutrition of plants is homogenous, controlled, and stable. Specific formulations developed for specific crops will maximize yields.
3. Plants can be spaced closer together, not limited by water or nutrient availability, only by available light. This results in higher yields per unit area the plant occupies.

4. No weeds, no cultivation.

5. Efficient use of water with automation of irrigation cycles. Water is managed to suit the specific stage of plant growth and crop.

6. Quality of fruit is firmer with longer shelf-life that leads to less shrinkage on supermarket shelves.

7. The use of fertilizers is efficient in that the nutrients are directed to plant roots uniformly and are readily available. No loss of nutrients occurs beyond the root zone.

8. The products are sanitary with no biological disease organisms present.

9. Transplant "shock" is minimized by use of growing cubes to retain roots in the cubes ready to grow out immediately after transplanting to the growing system.

10. Plants mature sooner with hydroponics as they are not under nutritional or water stresses.

11. Yields are at least 20% greater over soil culture unless you are a well-experienced soil grower, then such yields may be equal to those of hydroponics. However, under the extended growing season in a greenhouse, annual yields may exceed two or three times that of outdoor soil growing.

12. Constructing growing beds at waist height or the use of vertical plant towers will relieve you of back pains.

13. It is a very clean method of gardening, no messy hands from soil and its constituents.

14. There is no need to worry over the invasion of your indoor hydroponic garden by many of the troublesome animals outside looking for a meal, such as rabbits, deer, gophers, raccoons, woodchucks, moles, mice, and small rodents. If any small rodents get into your home and indoor garden, you can easily trap them. But, their entrance into your home should seldom, if at all, occur.

15. Most essential gardening tools for your outdoor vegetable garden are eliminated. No need for hand trowels, cultivators, hoes, shovels, garden forks, rakes, wagons, garden carts, power tillers, and so on.

With all of these benefits why not garden hydroponically? These benefits far outweigh the more precise procedures to follow for hydroponic growing. Besides, many small household units are designed to minimize your making mistakes. Nutrients are prepared, so you just add them to the water in amounts clearly set out in the directions. If you wish more challenges, you can derive your own nutrients from basic fertilizer salts. It is a very satisfying hobby that will uplift your spirit by producing healthy vegetables of superior flavor and nutrition.

3 Why You Should Garden Hydroponically

Caring for your plants in hydroponic gardening is similar in terms of their training, pollination, pest and disease control, and so on; however, it takes away the less-desirable tasks of weeding, cultivation of the soil, hand watering, and fertilizer applications, which is less back strain. Weeding is eliminated as is cultivation of the medium. Watering and fertilizer application is automated with flood systems or drip irrigation. There will be much less use of pesticides as most pests and diseases can be eliminated from the medium, whereas in soil they are always present attacking the plant roots (Figure 3.1). You will feel much better in seeing your plants healthy and productive. When pests attack the roots of plants, it is often very difficult to identify the causes of your plants suffering from wilting, yellowing, and often dying. Once you determine these causes of poor plant growth, the next thing is to treat the soil with some type of pesticide. This often involves using fairly strong synthetic pesticides that are not organic based. This creates apprehension coupled with caution in applying them safely. The outcome is that your plants are no longer free of synthetic pesticides ("organic"). Even then, due to the complexities of soil composition, the treatment may not control all of the pests and will have to be applied numerous times to maintain the pest populations at tolerable levels to minimize restrictions in plant growth.

Hydroponic gardening will produce more healthy plants in both safety and nutrition. Many soil-borne pests can be avoided in hydroponics. If an infestation should be introduced, control is more effective and usually done by organically derived pesticides known as "bioagents." Apply these natural pesticides to the growing medium through the nutrient solution with the drip irrigation system of hydroponic culture. The bioagents are much safer to handle than strong synthetic pesticides. These bioagents are the same ones recommended for certified "organic" growing. There is less need for using pesticides with hydroponic culture compared with soil gardening due to the use of relatively sterile substrates free of pests. Of course, the control of pests and diseases of the foliar part of the plants will be very similar to those growing in soil. However, with optimum nutrition, the plants will have thicker cuticles and stronger cell walls to help them resist infestation.

Optimum nutrition of hydroponically grown plants enables them to yield fruit (vegetables) higher in vitamins and minerals than those of their counterparts grown in soil. Soil is heterogeneous in structure, composition, and mineral content. Some plants may grow well in one area and within a short distance others may suffer from deficiencies. In hydroponics, the plants grow in a substrate homogeneous in water and minerals with pH levels maintained optimum for mineral uptake by the plants. In soil, it is more difficult to regulate pH at ideal levels to make elements available to plant-root uptake. This equal availability of essential elements to hydroponically grown crops gives superior quality and yields. That will give you greater satisfaction in your labors of gardening plus more production that you may share with others in your neighborhood (Figure 3.2). I think one of the most disappointing aspects of gardening is when you do everything you can possibly think of that should make your plants grow well and for some mysterious reason(s) they do not give you the results you are expecting. Avoid this by growing hydroponically. Initially, you may think that hydroponics is too technical, but, that is not the case. It is really important to understand how the plants grow and what their needs are as well as what kinds of problems can stress them to reduce their productivity. This applies to all gardening, not just hydroponics. Hydroponics is a science and by following the procedures you

FIGURE 3.1   Avoid spraying strong pesticides in a soil garden. (Drawing courtesy of George Barile, Accurate Art, Inc., Holbrook, New York.)

FIGURE 3.2   Productive hydroponic growing versus less-productive soil growing. (Drawing courtesy of George Barile, Accurate Art, Inc., Holbrook, New York.)

can be successful making your labors of growing very rewarding. With the presence of many hydroponic shops and online stores you may purchase readymade hydroponic units and all supplies such as nutrients, bioagents, and others for successful growing. In addition, there is lots of help by store operators and through the Internet on seeking solutions to any challenges that occur. With all of these considerations why not garden hydroponically?

Section II

*Understanding Hydroponics
and How Plants Grow*

4 Why Hydroponics Is Not Just Chemistry

Many people are of the opinion that hydroponics is all chemistry and that the plants grown by this technique are "inorganic" (Figure 4.1), which is not true. Some basic high-school level chemistry will help one understand how to prepare nutrient formulations, but even without such a background one can easily learn the procedures. Preparing nutrient solutions can also be done by purchasing ready-made nutrient mixtures. As far as the plants are concerned, they require the same essential elements regardless of whether they are obtained from hydroponic sources or by natural breakdown in the soil. The advantage of hydroponics is that one can provide the plants with optimum levels of each of the essential elements through the nutrient solution formulation (Figure 4.2). When growing in soil these same nutrients are added by the application of fertilizers and compost. However, because of the heterogeneity of the soil, it is more difficult to apply the nutrients at levels that are optimum for plant growth (Figure 4.3).

The nature and properties of the soil determine the availability of nutrients to plants. Different types of soils, such as, sand, sandy loam, loam and clay, are determined by their percentages of natural inorganic particle sizes and organic matter. Sandy soil composed of large mineral particles permits water and nutrients to move quickly through it and past the root zone of plants. These are not ideal for vegetable growing unless large amounts of water and fertilizers are supplied regularly. Pure igneous (volcanic) sand in fact is suitable as a hydroponic medium, where oxygenation to plant roots is readily available. At the other extreme is clayey soil that consists mainly of small particles that hold together tightly retaining water and minerals. This type of soil often has excess water with poor drainage causing lack of oxygen to plant roots. With this poor aeration, plants also suffer from lack of mineral uptake. A loamy soil has a good mixture of large and small minerals plus organic matter (humus), which provides adequate minerals, water, and oxygen to the plants. Maintaining soils in this optimum state of structure and fertility is often challenging, requiring soil tests and frequent additions of fertilizers and compost of adequate quantities for plant growth.

With hydroponics, the choice of substrate depends on the availability, cost, crop, water retention, oxygenation, structural integrity, and sterility. For most backyard gardeners, the availability and cost of the substrate are not restrictive because they use relatively small amounts. Some crops grow better in more porous substrates, whereas others grow well where there is higher water retention. However, oxygenation is important to all plants, so drainage is critical, especially for long-term crops such as tomatoes, peppers, eggplants, cucumbers, and other vine crops. Some short-term crops, such as herbs (basil, mint, and watercress) and lettuce can grow in water

FIGURE 4.1 Hydroponics is not specialized laboratory chemistry. (Drawing courtesy of George Barile, Accurate Art, Inc., Holbrook, New York.)

FIGURE 4.2 Use of scales for weighing and adding nutrients to a nutrient tank. (Drawing courtesy of George Barile, Accurate Art, Inc., Holbrook, New York.)

culture systems. Structural integrity, the ability of the substrate to retain its structure and not break down during the growth of the plants, is basic to hydroponic growing. This quality and sterility are of prime importance in the selection of a medium. If the substrate is not free of pest and disease organisms, they will attack the plant roots causing decreased plant vigor and yields. You will then be in a similar situation as what often occurs with soil growing. All of the variable properties of soils that can restrict plant growth through lack of oxygen and mineral availability, and/or occurrence of structural breakdown, and the presence of pests and diseases are difficult to

FIGURE 4.3 A gardener spreading fertilizers on the soil in his garden. (Drawing courtesy of George Barile, Accurate Art, Inc., Holbrook, New York.)

control. With hydroponics, you choose the best substrate that provides optimum levels of oxygen, minerals, and water. In addition, most of the pest and disease factors in a soilless substrate are avoided.

Overall, hydroponics and soil growing are not different with regard to the needs of the plants. The chemistry behind hydroponics is not different from soil growing with regard to providing ideal levels of nutrients to the plants. Only the procedures and some sources of nutrients differ. With soil we like to use slow-release compounds that will not rapidly pass beyond the roots of the plants, whereas in hydroponics we want highly soluble compounds that will dissolve completely in water because the nutrient solution is applied directly to the plant roots. The chemistry is the same for the plants as they must actively take up the same nutrients and water from the soil solution of the soil, or from the nutrient solution in hydroponic culture.

The principal difference between soil and hydroponic cultivation is this precise management of the availability of the essential elements to the plant roots under hydroponics. The other techniques in the care of the above-the-ground portion of the plants are similar in both hydroponic and soil cultures. All aspects of plant training, pest and disease control, even watering by a drip irrigation system also apply to soil culture. One step further is to extend your growing season by the control of environmental factors such as temperature, light, carbon dioxide, and relative humidity through greenhouse growing. You may grow either soil or soilless in the greenhouse, but normally it is advantageous to use hydroponics under controlled environments to maximize the health and yields of your plants.

5 Plant Growth
The Environment and Its Effects on Plants

Plants are composed of 80%–95% water. Plant dry matter is 10%–20% of the fresh weight. Over 90% of the dry matter of plants is composed of carbon (C), hydrogen (H), and oxygen (O). We are all familiar with the term photosynthesis. This is the process whereby light supplies energy, water from the growing medium provides hydrogen and oxygen, and carbon dioxide (CO_2) from air produces carbon and oxygen that become the building blocks (sugars) for plant growth (Figure 5.1). All of the other elements needed in photosynthesis, making up 1.5% of the fresh weight, are from the soil or nutrient solution. These are the essential elements that we discuss in Section III of this book.

Photosynthesis may be expressed as an equation as follows:

$$6CO_2 + 6H_2O \xrightarrow{\text{Light}} C_6H_{12}O_6 + 6O_2$$

$$\text{Carbon dioxide} \qquad \text{Water} \qquad \text{Sugar} \qquad \text{Oxygen}$$

The sugar is a form of chemical energy that is used to drive all the plant's processes. Plants are the basis of almost all life on our planet and photosynthesis the source of energy for nearly all life on Earth. Photosynthesis uses light visible to our eyes (Figure 5.2). The light is absorbed by chlorophyll, the green pigment, in all plant parts, especially in the leaves where most organelles called chloroplasts are located. The chloroplasts contain chlorophyll-*a*, chlorophyll-*b*, and carotenoid pigments. Most absorption of light is in the violet–blue and red light of the visible spectrum as shown by the absorption spectrum of these pigments (Figure 5.3). When we use supplementary lights for our plants indoors, we want light that gives off most energy in this part of the visible light. There are many complex processes that take place within the plant to convert the sugar into carbohydrate products by carbon fixation whereby carbon is taken from sugars and combined to form sucrose and starch. The carbon from photosynthesis is used to form other organic compounds such as cellulose, lipids, and amino acids or others to fuel respiration.

In respiration, metabolic reactions take place in the cells of plants (and animals) to convert biochemical energy from nutrients into high-energy molecules that can later break down into smaller molecules releasing energy in the process. Respiration provides the energy to fuel cellular activity. The nutrients used by animal and plant

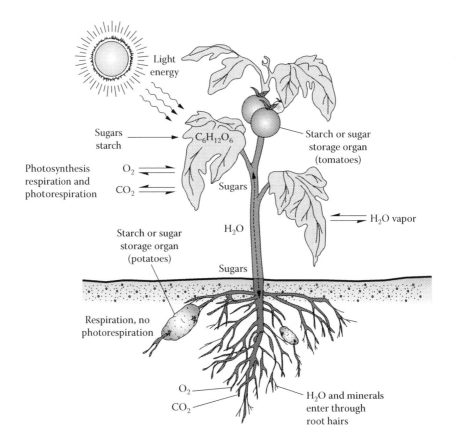

FIGURE 5.1 Photosynthesis process in plants with movement of water and manufactured sugars, and so on flowing to the roots and fruits. (Drawing courtesy of George Barile, Accurate Art, Inc., Holbrook, New York.)

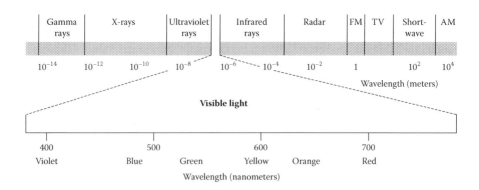

FIGURE 5.2 Visible light spectrum. (Drawing courtesy of George Barile, Accurate Art, Inc., Holbrook, New York.)

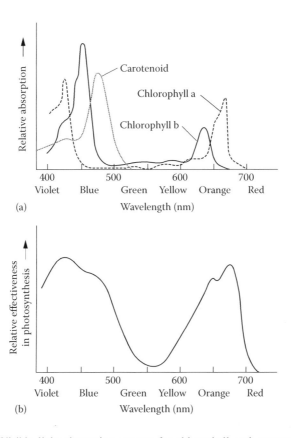

(a)

(b)

FIGURE 5.3 Visible light absorption spectra for chlorophyll and carotenoid plant pigments (a) and photosynthetically active radiation (PAR) (b). (Drawing courtesy of George Barile, Accurate Art, Inc., Holbrook, New York.)

cells in respiration include sugar, amino acids, and fatty acids. The energy is stored in the high-energy molecule adenosine triphosphate (ATP) and during oxidation (use of molecular oxygen), the energy stored in ATP is released to drive energy processes such as biosynthesis, locomotion (movement in animals), or transportation of molecules across cell membranes.

A simplified reaction for respiration is as follows:

$$C_6H_{12}O_6 \quad + \quad 6O_2 \quad \rightarrow \quad 6CO_2 \quad + \quad 6H_2O \quad + \quad Heat$$

| Sugar | Oxygen | Carbon dioxide | Water |

Because respiration requires oxygen in plants, it is termed as aerobic respiration. It is the main process by which both fungi and plants break down organic compounds into energy needed for their growth. These organic compounds are produced during photosynthesis. In plants, respiration occurs during the dark. Consequently, at night the plants use oxygen and give off CO_2 and water.

It is important to understand these simplified basics of plant growth in order to know the key factors of the environment that plants require for healthy development. When these factors are not at optimum levels, they will limit plant growth and therefore can be termed limiting factors. During the day, plants need the correct quality and intensity of light to drive photosynthesis. The quality refers to the color of light (determined by its wavelength). Plants require light between 400 and 700 nm wavelength, which is in the visible spectrum as shown in Figure 5.2. A nanometer is 10^{-9} or 1/1,000,000,000 m (one billionth of a meter) in length. This light source that plants utilize is termed photosynthetically active radiation (PAR). This designates the solar radiation from 400 to 700 nm that plants actively utilize in photosynthesis (Figure 5.3).

Increasing light energy in the PAR range increases photosynthesis. Each crop has an optimum light intensity that maximizes plant growth. If there is insufficient light, plant growth slows down and if excess light is given, plant growth will not increase (Figure 5.4). As a result, when using lights you must be sure to give sufficient, but not excess as the cost of the additional light will not result in increased production. The quantity of light is the intensity that can be measured. In the United States, the unit for measuring light intensity is the foot-candle, whereas lux is used in Europe. An argument against the use of foot-candles is that it primarily measures visible light detected by the human eye and not necessarily the amount of light a plant receives. Most horticulturists use a unit that measures light at any instant in micromoles (μmol) per square meter (m^{-2}) per second (s^{-1}) of PAR. This unit measures the number of photons (individual particles of energy) used in photosynthesis that fall on a square meter of surface every second. Because this is an instant reading, the better unit to use is the daily light integral (DLI), which is the amount of PAR received each day (moles per day). In greenhouses, the values are normally less than 25 mol/m/day. To grow plants in your home using artificial lights you need to get sufficient light for optimum yield. Researchers have

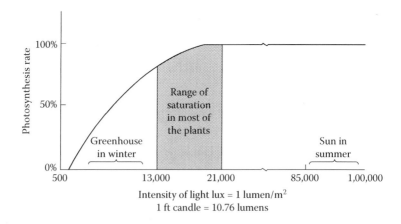

FIGURE 5.4 Graph of photosynthesis activity versus light energy (intensity). (Drawing courtesy of George Barile, Accurate Art, Inc., Holbrook, New York.)

developed DLI levels for groups of plants classifying them as low-light, medium-light, high-light, and very-high-light crops. Fruit-bearing crops such as tomatoes, peppers, and European cucumbers would lie in the very-high-light crops. Although this is a little more technical, it shows you what amount and type of light is best for your crops.

Another environmental factor that affects plant growth and yields is temperature (Figure 5.5). Again all crops have different optimum temperature ranges. Crops are divided into cool-season crops and warm-season crops. Cool-season crops include cabbage, cauliflower, broccoli, and lettuce, whereas warm-season crops include fruiting crops such as tomatoes, peppers, cucumbers, and eggplants. Normally, cool-season crops require night temperatures in the 50s F (10–15°C) and 60s F (16–21°C) to low 70s (22–23°C) during the day, whereas warm-season crops like 65°F (18°C) or higher at night and 75–80°F (24–27°C) during the day. When you browse through seed catalogs searching for varieties of crops to grow, information will be available on their optimum temperatures. If not, simply look up on the Internet search engines for crops and their ideal temperatures. Of course, temperature can only be regulated in greenhouse or indoor gardening, not outside in prevailing weather conditions. This, however, is significant with hydroponic growing as in most cases hydroponic culture is most applicable to greenhouse or indoor growing.

Under very-high temperatures and especially with low relative humidity (RH) (percentage of moisture in the air) plants will slowdown in growth due to their inability to keep their tissues at optimum temperatures. This causes the closing of stomata (small pores particularly numerous on the lower sides of leaves) to partially or fully close. The closing of the stomata blocks the entrance of CO_2 into the leaves and restricts water loss that in effect reduces cooling of the plants through evapotranspiration (loss of water by evaporation and transpiration). It will then reduce water and resultant nutrient uptake slowing growth further. As was pointed out earlier, plants receive CO_2 from the air as part of the photosynthesis process. Any environmental factors that are not at optimum levels for the specific crops, will restrict photosynthesis and subsequent plant growth and development (Figure 5.6). When these environmental factors are restricting or limiting growth, they are termed "limiting factors."

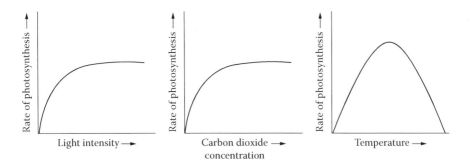

FIGURE 5.5 Graph of photosynthesis activity versus light intensity, carbon dioxide, and temperature. (Drawing courtesy of George Barile, Accurate Art, Inc., Holbrook, New York.)

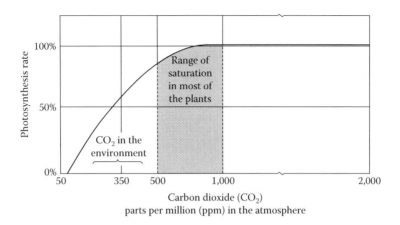

FIGURE 5.6   Graph of photosynthesis activity versus carbon dioxide. (Drawing courtesy of George Barile, Accurate Art, Inc., Holbrook, New York.)

With hydroponic gardening in greenhouses and indoors, you must be aware of the optimum levels of light, temperature, CO_2, and RH for your crops, monitor and regulate them at levels best for crop growth to maximize yields. This is discussed in more detail in Section V under greenhouse and indoor growing.

6 Water Needs, Management, and Irrigation Practices

Water is essential for all life forms, including plants. Plants use more water than animals. Plants are 90% water, whereas animals may be 75% by weight. As mentioned in the previous chapter, water is part of the photosynthesis process and is present in cells. From the very beginning of seed germination, water is essential. Water is the medium by which plants take up minerals from the soil solution or the nutrient solution in the case of hydroponics. Plant roots actively absorb the nutrients from the water and transport all chemicals in and out of cells by water. The water is absorbed into the plant near the tips of the root by specialized root hair cells (Figure 6.1). From the root hairs the water must enter the vascular tissue (xylem) that transports the water throughout the plant (Figure 6.2). This is done through a selectively permeable membrane, a single layer of cells called the endodermis. This movement of water into the endodermis is the water going from a region where it is at a higher concentration to one where its concentration is lower (in the cell). This process is termed "osmosis."

Water moves upward in plants through the xylem cells, which are long, narrow, tubes containing no living matter (Figure 6.2). They are joined end-to-end to create long tubular pathways from the roots through the stem to the leaves. The water moves up not by just capillary force, but by the cohesion force of water molecules. Water is lost from leaves by evaporation through the leaf stomata (Figure 6.3). This is transpiration, also termed evapotranspiration, whereby the water moves out of the leaves and is the driving force to pull the water through the plant in its xylem vessels. In most plants, more than 95% of the water taken in by the roots is lost through evapotranspiration in the leaves. This has a cooling effect on the plant tissues. Higher temperatures and increased wind speeds increase the transpiration rate. As the transpiration increases, the uptake of water by the plant roots must increase to keep the plant turgid. If water uptake is less than water loss, the deficit in the plant will cause the stomata to close and the plant will lose turgidity causing wilting of leaves and then stems. This kind of stress will slow growth and production as when the stomata close, carbon dioxide cannot enter either, so the whole process of photosynthesis slows down or stops if water is not made available to the plant. This occurs in soil when it dries out to a level that the plants cannot take up sufficient water. It can also happen in hydroponic culture if there are large intervals between irrigation cycles and the substrate has insufficient water available to the plant.

In many fruiting crops, such as tomatoes and peppers, a water deficit in the plant will result in blossom-end rot of the fruit. This is caused by insufficient water uptake

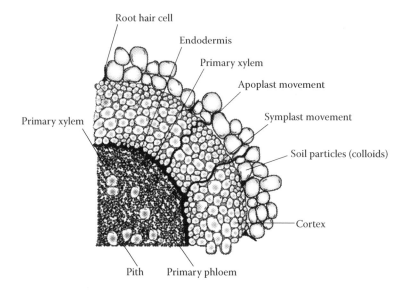

FIGURE 6.1 A cross-section of a root with movement of water and minerals. (Drawing courtesy of George Barile, Accurate Art, Inc., Holbrook, New York.)

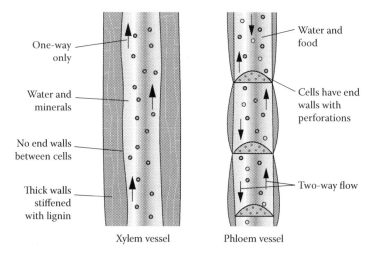

FIGURE 6.2 Xylem and phloem conducting vessels. (Drawing courtesy of George Barile, Accurate Art, Inc., Holbrook, New York.)

and resultant loss of calcium uptake. The symptom is a dry, leathery-like, black tissue at the blossom end of the fruit. High humidity reduces transpiration rates, low humidity accelerates transpiration. An example of the effect of relative humidity on production is given by lettuce. Under high relative humidity, the plant does not release adequate water, so this slowdown of water movement from the root to the leaves causes a lack in calcium uptake resulting in "tip burn" (blackening of leaf margins) of lettuce. If you understand these functions of water within the plant and

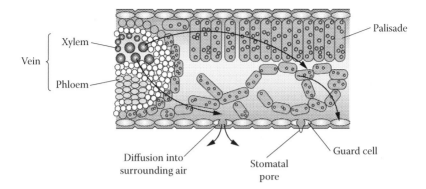

FIGURE 6.3   Plant upward movement of water and minerals in the xylem with water leaving the leaf through the stomata during the evapotranspiration process. (Drawing courtesy of George Barile, Accurate Art, Inc., Holbrook, New York.)

how its lack can cause stress in the plant leading to symptoms and reductions in yields, you will know what signs to watch for and know when you must add water or reduce it through the irrigation cycles.

The nutrient transport system is the function of the phloem tissue (Figure 6.2). It, like the xylem, requires water as a medium to transport the photosynthetic products (photosynthates) throughout the plant from its source to the areas of utilization (sinks) (Figure 6.4). The sinks include all areas of the plant—roots, stems, and fruit—to where these food substances are transported. If you permit fruit such as tomatoes or peppers to ripen completely on the plant, you will get better flavor and higher nutrition in the fruit harvested "vine ripened" than picking the fruit before it is fully ripened. This is the outcome of allowing the fruit ("sink") to accumulate more of the food substances as it matures.

Water management is controlling the amount of water supplied to the soil or hydroponic substrate to get optimum growth by avoiding any stresses to the plant. With hydroponics you will generally use an automated system. Irrigation controllers enable the gardener to set irrigation cycle frequency according to the plant stage of growth and weather conditions. You will irrigate more frequently and with longer duration of any cycle determined by light, temperature, relative humidity, and day length. The nature of the growing substrate influences the irrigation practices. More coarse particles will require more frequent irrigation cycles than finer particles. For example, perlite substrate may need five to six cycles per day, whereas coco coir or a peatlite medium that has higher water retention, two to three daily cycles would be adequate. The principle is to keep the levels of nutrients sufficient to be readily available to the plant roots at all times. However, excessive cycles can cause lack of oxygen due to too much free water in the void spaces of the substrate. Oxygen is critical to the plant roots to allow active transport of elements into the plant.

When irrigating plants to keep the nutrient solution from concentrating in the medium by evaporation, it is also important to have a percentage of leachate to occur during irrigation cycles. This will be a function of the length of time the irrigation is activated during any cycle. The percentage of leachate varies with the substrate.

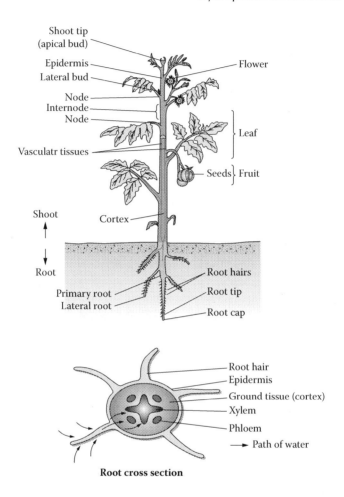

Shoot tip
(apical bud)

Epidermis

Lateral bud

Node

Internode

Node

Vasculatr tissues

Flower

Leaf

Seeds } Fruit

Shoot

Root

Cortex

Root hairs

Root tip

Root cap

Primary root

Lateral root

Root hair

Epidermis

Ground tissue (cortex)

Xylem

Phloem

→ Path of water

Root cross section

FIGURE 6.4 Movement of water and nutrients in the xylem and manufactured photosyn-thates in the leaves flowing through the plant stem, and so on. (Drawing courtesy of George Barile, Accurate Art, Inc., Holbrook, New York.)

In general, with rockwool and perlite cultures we want approximately 25% leachate, whereas with coco coir and a peatlite medium it should be approximately 10%–15%. When growing in soil, gardeners can use a tensiometer that tests the moisture level in the soil. With hydroponics, moisture sensors may be placed in the medium that sense the moisture level and will activate an irrigation cycle automatically based on a preset limit. This kind of feedback system provides the crop with more uniform irrigation than a simple time-clock type of controller.

In summary, recognizing the factors that determine the water usage by plants assist you in managing the irrigation practices to keep plants most productive. Water quality, mineral content, and plant consumption under variations of weather and plant growth, plant appearance, and symptoms all assist in recognizing any imbal-ances with irrigation of the plants.

Section III

**Nutrients Essential to Plants
and Their Sources**

7 Essential Nutrients to Plants and Their Functions

Minerals that are required for plant growth and development are termed "essential." These include carbon (C), hydrogen (H), and oxygen (O) as we described earlier are part of the photosynthesis process. These come from the air and water. The remaining essential elements come from the soil or nutrient solution in the case of hydroponic culture. These include nitrogen (N), phosphorus (P), potassium (K), sulfur (S), calcium (Ca), and magnesium (Mg), which are required in relatively large amounts and therefore termed macro- or major elements. The others needed in very small amounts are termed micro-, minor, or trace elements. These include iron (Fe), manganese (Mn), zinc (Zn), copper (Cu), boron (B), molybdenum (Mo), and chlorine (Cl). Nickel (Ni) is now believed to also be an essential element. Other elements accumulate in some plants and may be used in their growth. These are silicone (Si), aluminum (Al), cobalt (Co), vanadium (V), selenium (Se), and platinum (Pt). However, when we speak of the elements that we must provide for our plants, whether in soil or hydroponics, they are the six macro- and seven microelements (eight if we include nickel) listed above.

Each of these essential elements has specific functions within the plant. It is helpful if we understand what these functions are to assist us in recognizing nutritional disorders that may occur in the plants.

Nitrogen—Part of organic compounds, including proteins, nucleic acids, and chlorophyll.

Phosphorus—Plays a role in respiration and cellular division and is used in the synthesis of energy compounds—adenosine triphosphate and adenosine diphosphate.

Potassium—Usually found in the meristems (tips of plants) where it activates many enzymes.

Calcium—Vital part of cell walls holding them together, maintains membrane integrity and acts in the movement of substances through cell membranes.

Magnesium—Essential component of the chlorophyll molecule and activates many enzymes.

Sulfur—Part of many amino acids.

Iron—Essential for chlorophyll synthesis, enzyme activator, and acts as an electron carrier in photosynthesis and respiration.

Manganese—Enzyme activator.

Boron—Involved in calcium ion use.

Zinc—Enzyme activator.
Copper—Acts as an electron carrier and is part of certain enzymes.
Molybdenum—An electron carrier in conversion of nitrate to ammonium.
Chlorine—Acts as an enzyme activator in photosynthesis.
Nickel—Essential for urease enzyme activity.

If any of these elements is in deficiency or excess, disorders will occur in the plants. These disorders will be expressed as symptoms. Symptoms (specific colors or deformities) will give you a clue that your plants are under stress and must be corrected to avoid loss in production.

The ability of the soil or hydroponics to provide adequate nutrition through the availability of the essential elements to plant roots depends on the amounts of the various elements present, their solubility (presence in the soil water or nutrient solution in a solution and not just a suspension), and the pH of the soil or nutrient solution. Soil nutrients exist in complex, insoluble compounds, and soluble forms readily available to plants. In hydroponics, highly soluble compounds are dissolved in water to obtain the nutrient solution that has the elements readily available to the plants. The reaction of the soil or hydroponic solution (pH) determines the availability of the various elements to the plant (Figure 7.1). The pH is a measure of the acidity

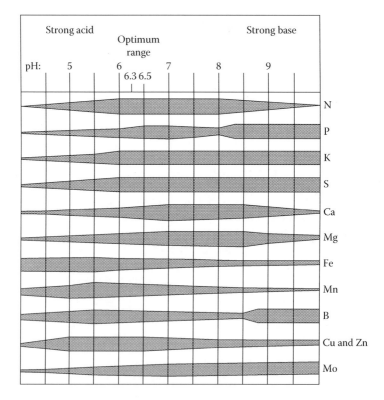

FIGURE 7.1   The effect of pH on the availability of plant nutrient uptake. (Drawing courtesy of George Barile, Accurate Art, Inc., Holbrook, New York.)

or alkalinity. If the solution pH is less than seven it is acidic, seven is neutral, and greater than seven is alkaline. Most plants prefer a pH between 6.0 and 7.0 for optimum nutrient uptake regardless of whether it is the soil solution or a nutrient solution. Specific crops require different optimum pH ranges. For example, lettuce likes a pH between 5.5 and 5.8, whereas tomatoes, peppers, and cucumbers prefer a pH from 6.0 to 6.4.

If you encounter plant symptoms pointing to a lack of a specific element, be sure to check the pH of the solution in case the element may be present in ample amount, but, unavailable to the plant due to the incorrect pH. The pH can be tested with indicator papers, dye solutions, and pH meters. In hydroponic systems, the pH must be tested at least once a day. If it is too low, raise it with a base such as potassium hydroxide, or on a small-scale use bicarbonate of soda. With high pH values lower the pH with an acid such as sulfuric (battery) acid or phosphoric acid. You could also use vinegar (acetic acid) or citric acid for home use. The pH of cider vinegar at normal strength is 4.25–5.00. White vinegar is stronger with approximately 5%–8% acetic acid in water. It has a pH of approximately 2.4. An even easier method is to purchase a "pH Up" or "pH Down" solution from a hydroponic retail outlet or from an Internet website of hydroponic suppliers. Always remember to add acid to water to avoid splashing or fumes. It is best to use eye protection and wear gloves when using strong acids or bases to avoid burns.

8 Sources of Nutrients for Plants in Soil versus Hydroponics

Although the elements essential for plants are the same, regardless of whether they are from sources in the soil or the nutrient solution, their original form may differ. In soil, these elements come from the break down of organic matter through microbial organisms and animal decomposition. Animals would be, for example, earthworms that consume organic matter of the humus of the soil and excrete simpler forms of organic compounds. These compounds are further decomposed into more simple compounds by the microorganisms. The end process is the release of inorganic elements into the soil water to form the soil solution that brings these essential minerals in contact with plant roots where they are absorbed. The elements must be in their charged atomic state (ions) to be taken up by plants. The other component of soil that is a source of minerals is the sand, silt, clay, and rocks that break down through weathering. Wind and water will break them into very fine particles that when in contact with the soil water will be released as ions, once again available to the plant roots.

In adding nutrients to soil, we supplement with composts, manures, peatlite mixes, perlite, fertilizers, and so on to also improve the structure of the soil (Figure 8.1). Generally, with soil growing the choice of fertilizer depends on the plant and the results of a soil analysis. Normally, blends are used. Fertilizers have a guaranteed analysis that appears on their bags. All chemical and organic fertilizers have their guaranteed analysis on their label. For example, a common vegetable garden fertilizer is 5–10–5. This fertilizer contains 5 percent nitrogen, 10 percent phosphate (P_2O_5), and 5 percent potash (K_2O). The particular fertilizer should be chosen after a soil analysis. The laboratory doing the analysis will make recommendations as to the fertilizer components needed for your soil. The pH of the soil may be modified by the use of lime to raise it or sulfur to lower it. Most fertilizers used in soil gardening are available in granular and water-soluble forms. Granular fertilizers slowly release the plant nutrients to the soil water and therefore act slowly, but have the advantage of being long lasting. Water-soluble fertilizers are fast acting, but, move rapidly through the soil, so must be applied more frequently than granular forms.

In hydroponic culture, we grow in containers or some form of medium that is wrapped with plastic, such as "slabs" that are long, narrow, bags with the substrate. They generally measure 3 ft by 6–8″ wide by 4″ thick. These slabs will sit on return channels underneath that will collect the spent solution (leachate) and recycle it or drain it away from the growing area (Figure 8.2). For this reason, because the nutrient

FIGURE 8.1   Gardener adding fertilizers to soil bed not knowing exactly how much is required. (Drawing courtesy of George Barile, Accurate Art, Inc., Holbrook, New York.)

Re-circulation of nutrient solution in hydroponics

Black poly lateral

Inlet header

Drip line

Rockwool block

Slab

Collection tray

Leachate from slab

Drain/return pipe

Nutrient tank

Pump

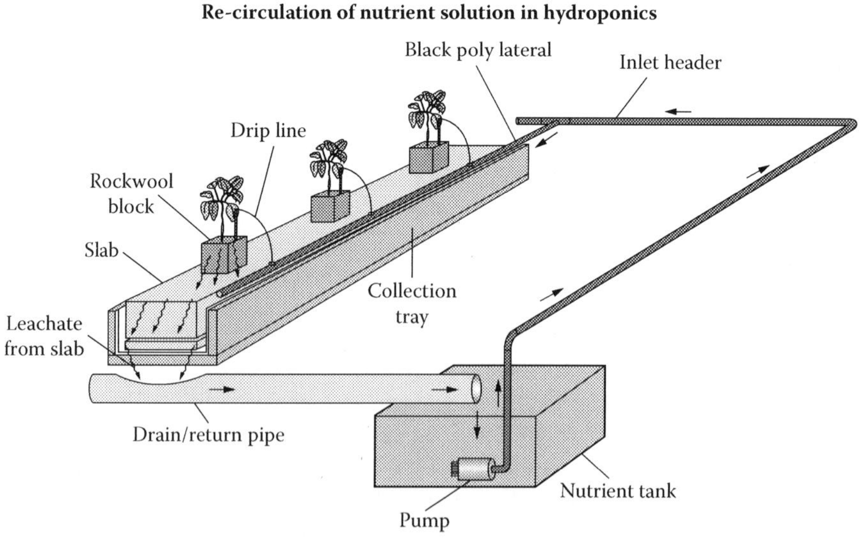

FIGURE 8.2   Nutrient solution is recycled at the root zone in a closed hydroponic system. (Drawing courtesy of George Barile, Accurate Art, Inc., Holbrook, New York.)

soluton does not just pass the root zone but is distributed precisely near the base of the plants with a drip irrigation system highly pure and soluble fertilizers are used. These highly soluble fertilizers insure that all of their elements are released to the water to form the nutrient solution. The nutrient solution is complete in containing all 13 essential elements in the correct concentratons, measured in parts per million (ppm) or milligrams per liter (mg/L), for optimum plant growth.

The sources of the plant nutrients must be highly soluble and of high purity. Hydroponic suppliers handle many blends of nutrient solution so that you may choose the best for your crop (Figure 8.3). There are different formulations for different stages of plant growth, starting solution, initial vegetative growth, flowering stage, and fruit production. There are vegetable formulations as well as flower and ornamental ones.

Most prepared nutrients come in two components; "A" and "B." A few are available as just one mixture. The use of a two part formulation is usually better to prevent any possible reaction from occuring among the various elemental components. Usually one will contain calcium, nitrogen, potassium, and iron. The other will have the rest of the elements, including the microelements. They are at concentrated levels when packaged or bottled, so cannot be mixed at those levels or they will react to form an insoluble hard substance, a precipitate. The precipitate cannot be re-dissolved in water. For example, if you mix concentrated calcium or iron with a sulfate, such as Epsom salts (magnesium sulfate) precipitation will result in an insoluble form of calcium sulfate or iron sulfate. Dissolve parts A and B separately in water to prevent any reaction. The ingredients of part A are normally calcium nitrate, potassium nitrate, and iron chelate. Part B may contain potassium nitrate, potassium sulfate, monopotassium phosphate, magnesium sulfate, and other sulfates of manganese, zinc, and copper. In addition, part B will have the remaining trace elements of boron, molybdenum, and chlorine.

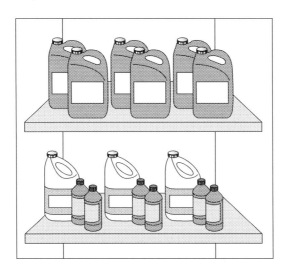

FIGURE 8.3   Many different packaged nutrient formulations available at hydroponic shops. (Courtesy of Botanicare, Tempe, Arizona.)

If you wish to take on the challenge of making up your own nutrient formulations, that is discussed in the following chapter. As I mentioned above, the compounds you select to add the essential elements for the nutrient solution must be pure and highly soluble. The following is a list of recommended compounds for the solution makeup.

Calcium Nitrate: Molecular Formula: $Ca(NO_3)_2$
    It is important to purchase "Greenhouse Grade" to avoid the presence of a greasy plasticizer on lower grades. One source is called "YaraLiva" CALCINIT™ Greenhouse Grade, 15.5–0–0 with 19 percent calcium. It is a product of Norway.
    Another brand of highly soluble calcium nitrate is "Haifa Cal GG." This is also a greenhouse grade made by Haifa Chemicals Ltd., Haifa, Israel, and manufactured in Slovakia.
    This compound provides calcium (Ca) and nitrogen (N) for your plants.
Potassium Nitrate: Molecular Formula: KNO_3
    Once again select a soluble grade of fine powder form. "Yara Krista K" is a brand of soluble potassium nitrate 13.7–0–46 coming from Chile by SQM Industrial SA.
    This compound provides potassium (K) and nitrogen (N).
Magnesium Sulfate: Molecular Formula: $MgSO_4 \cdot 7H_2O$
    Highly soluble brands are a white crystalline powder; that is the heptahydrate (it has seven molecules of water attached as shown in the chemical formula). It is commonly known as "Epsom salts." It contains 9.8 percent magnesium (Mg) and 12.9 percent sulfur (S). PQ Corporation in Pennsylvania produces it. This can also be purchased in small quantities, very pure form at pharmacies.
Monopotassium Phosphate: Molecular Formula: KH_2PO_4
    A good highly soluble brand is Haifa Chemicals Ltd in Israel. The guaranteed analysis is 0–52–34. This compound gives potassium (K) and phosphorous (P).
Potassium Sulfate (Sulfate of Potash): Molecular Formula: K_2SO_4
    "Champion" water soluble grade ("crystalline") is produced by SQM in the United States. Its analysis is 0–0–51–17 where the last figure indicates 17 percent sulfur.
    It supplies potassium and sulfur for the plants.
Iron Chelate: Molecular Name: Sodium Ferric Diethylenetriamine Pentaacetate (Fe-DTPA)
    This is termed "Sprint 330" and made by Becker Underwood, Inc., in the United States. It has 10 percent elemental iron (Fe).
    Iron is a minor element, but is required in greater amounts than other minor elements. The plant needs may vary from 2.0 to 5.0 ppm. Other minor elements are at optimum levels less than one part per million.
Manganese Sulfate ($MnSO_4 \cdot 4H_2O$) or Manganese Chelate (MnEDTA).
    Either of these compounds is a source of manganese. If your water is alkaline, the chelate is a better source because the chelating agent will keep the manganese available to the plant.

Zinc Sulfate: Molecular Formula: $ZnSO_4 \cdot 7H_2O$
  This product is very soluble if the white powder form is used. It adds zinc (Zn) to the nutrient solution.
Copper Sulfate: Molecular Formula: $CuSO_4 \cdot 5H_2O$
  This is also termed bluestone due to its blue crystals. It is highly soluble in providing copper (Cu).
Boric Acid: Molecular Formula: H_3BO_3
  Boric acid, also called Boracic acid, is used as an antiseptic for minor burns or cuts and as an eye rinse. This provides boron (B). Boric acid is available at a pharmacy.
Ammonium Molybdate: Molecular Formula: $(NH_4)_6Mo_7O_{24}$
Sodium Molybdate: Molecular Formula: Na_2MoO_4
  Either of these is highly soluble in supplying molybdenum (Mo).

Note: With the minor elements, except iron, use very small amounts so purchase them as laboratory reagents in small quantities at most laboratory suppliers. They will be very pure and soluble.

Calculations to develop a nutrient formulation and how to make a nutrient solution with the formulation is discussed in Chapter 9.

As I mentioned earlier, you do not need to get this involved in making up your own formulation and storing all of these compounds. It is far simpler to purchase a ready-made formulation from a hydroponic supplier. However, as with most hobbies, you may wish to explore more technical details of growing your crops and experimenting with formulations to find the most optimum to maximize yields. Just enjoy the success of growing hydroponically at a level that suits you best!

9 Nutrient Formulations and Solutions

Specific crops have different optimum levels of each nutrient. These levels are measured in milligrams per liter (mg/L) or parts per million (ppm). One part per million is one part of one substance in a million parts of another. Water is the solvent in hydroponics as it is in soil for the elements (solutes). An optimum formulation depends on a number of factors.

Plant: Different plants like different levels of the essential elements.

Stage of Plant Growth: When plants are young developing seedlings, they need lower levels of macroelements. As they mature and start forming fruit some elements, such as potassium, calcium, and iron will be in more demand by the plant. When plants are growing, initially we add more phosphorous to promote root growth. As fruiting crops of tomatoes, peppers, and eggplants develop flowers and fruit, we can help the plant shift into a more generative flowering–fruiting phase from its initial rapid, leafy growth of a vegetative state. This can be done by the overall concentration of the nutrient solution, frequency and duration of irrigation cycles, and temperature control. With hydroponics we can observe these characteristics and assist the plant to be more productive by altering these factors, especially when growing indoors or in greenhouses.

Weather: Light levels and day length have a great effect on plant growth. With indoor and greenhouse gardening, we can supplement with artificial lighting, although that still is not as efficient as the natural sunlight. During short days of lower light, we can slow plant growth by adjusting the irrigation cycles, temperatures, and nutrient formulation.

Total dissolved solutes instruments measure the solutes in water by electrical conductivity (EC) (Figure 9.1). When solutes are dissolved in water, the solution will conduct electricity. The quantity and nature of solutes determines its EC. This is expressed as millimhos per unit volume (mMho). By monitoring the EC of a nutrient solution, one can determine when changes occur and know when to add elements or change the nutrient solution. Nutrient solutions having adequate essential elements possess an EC range between 1.5 and 2.5 mMho or slightly higher.

Plants take up much more water than nutrients, and at a greater rate. The volume of solution should be maintained relatively constant. That can be done by the use of an automatic float valve.

The total concentration of the nutrient solution elements should be between 1000 and 1500 ppm to facilitate uptake by the plant roots. Conductivity readings of these concentrations would correspond to 1.5 and 3.5 mMhos. Cucumbers prefer lower values (1.5–2.0 mMho), whereas tomatoes do better at higher values (2.5–3.5 mMho).

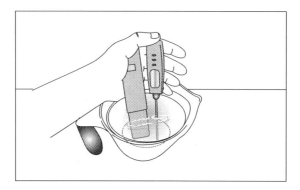

FIGURE 9.1   Person testing pH and electrical conductivity of the nutrient solution of hydroponic culture. (Drawing courtesy of George Barile, Accurate Art, Inc., Holbrook, New York.)

Overall, plants harvested for their leaves (lettuce, herbs) prefer high nitrogen levels because it promotes vegetative growth. On the other hand, fruit-bearing crops should have lower N and higher P, K, and Ca levels.

You may find many nutrient formulations online or in books, such as my book *Hydroponic Food Production*. Formulae for different crops are available from these sources. The following is a general formulation of macronutrients plus iron for a 20-U.S.-gal tank. Because weights are small, use grams instead of ounces or pounds. You will need a gram scale that can weigh accurately within 0.1 g.

MACROELEMENT FORMULATION

- Calcium (Ca):   180 ppm   Nitrogen (N):   140 ppm
- Phosphorus (P):   50 ppm   Potassium (K):   352 ppm
- Magnesium (Mg):   50 ppm   Sulfur (S):   168 ppm
- Iron (Fe):   5 ppm

WEIGHT/20-U.S.-GALLON TANK (GRAMS)

- Calcium Nitrate: 62 g   Potassium Nitrate: 8 g
- Potassium Sulfate: 46 g   Magnesium Sulfate: 38 g
- Monopotassium Phosphate: 17 g   Iron Chelate (10% Fe): 4 g

If you wish to use a larger or smaller volume, just use a ratio as a factor to multiply the weight. For example, if you want to make up only 10-U.S. gallons, multiply each weight of compound by the factor: 10/20 = 0.5. So, for calcium nitrate it would be 0.5 × 62 g = 31 g.

With the micronutrients because their weight is very small, you can make up a concentrated "stock solution" and then use a small volume of it to add to the 20-gallon tank. Use a 5- or 10-gallon water tank to store the micronutrient stock solution. Keep it in the dark to prevent algae growth. In this case, make up a 300 times normal strength stock solution as outlined for a 10-gallon volume.

MICROELEMENT FORMULATION

- Manganese (Mn):   0.8 ppm   Copper (Cu):   0.07 ppm
- Zinc (Zn):   0.2 ppm   Boron (B):   0.3 ppm
- Molybdenum (Mo): 0.03 ppm

WEIGHT/10-U.S.-GALLON TANK (300 TIMES NORMAL STRENGTH) (GRAMS)

- Manganese Sulfate: 41 g   Copper Sulfate: 3.2 g
- Zinc Sulfate: 11 g   Boric Acid: 20.5 g
- Ammonium Molybdate: 0.6 g

Now add a portion of this stock solution to the nutrient solution. Once again for using a 20-U.S.-gal tank add 20 × (1/300) = 0.066 U.S. gallons of the micronutrient stock solution to a 20-U.S.-gal tank. The factor 1/300 is to dilute the concentrate from 300 times back to normal one-time strength. Once again it is better to measure this small volume using milliliters (1 mL = 1/1000 L). The conversion to liters from U.S. gallons is 3.785 L per gallon. The conversion is 0.066 × 3.785 = 0.250 L or 250 mL. It is best to measure this volume with a 100-mL graduated cylinder. Scales and graduated cylinders should be available at a hydroponics shop or if not go online to a science laboratory supply distributor.

Now test the pH of the nutrient solution in the stock tank and adjust it either up or down using an acid or base as described in Chapter 7. Of course, the easiest method is to purchase a "pH Up" or "pH Down" solution from a hydroponic store (Figure 9.2). Add a small volume of the pH adjuster solution slowly while stirring to get good mixing. Check the pH with an indicator paper or pH meter as mentioned also in Chapter 7. Do not be afraid of exceeding the desired pH value as you can always adjust it in the opposite direction using the opposite solution from which you were using. If that happens, you are adding too much at any time between checking it.

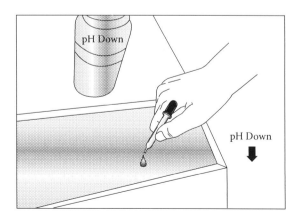

FIGURE 9.2 Person adding pH adjuster solution to nutrient tank. (Drawing courtesy of George Barile, Accurate Art, Inc., Holbrook, New York.)

Making up your own nutrient formulation is more involved than just purchasing a ready-made concentrate solution, but it is more fun. With commercial stock solutions, simply follow directions. They often tell you to use a number of teaspoons of their concentrate to each gallon of the tank solution. To be more accurate, use a small graduated cylinder. Have fun, this goes back to your school chemistry classes!

10 Signs of Plant Nutritional and Physiological Disorders and Their Remedies

Plants are similar to us humans and animals in that when under stress from poor nutrition; our bodies suffer in growth, development, and general health. Animals show these disorders in the form of weak bones, skin discoloration, and poor weight. Plants show nutritional defects in vigor, strength of stems, color of leaves, and poor yields.

Whenever plants undergo any type of stress from environmental conditions to lack or excess of nutrients, they will express signs of disorders. Pests and diseases also cause stress and disorders within the plant. Pest and disease causes and their control are discussed later in Chapter 25. Focus now is on plant symptoms from nutrient stress. By recognizing and segregating out what is the cause of a symptom, adjustments can be made in the environment or nutrient solution to remedy the stress and bring the plant back to healthy growth. In my book, *Hydroponic Food Production*, there is detailed information on nutritional and environmental effects on plants, how to determine the specific disorder, and the function of elements within the plant. Presented here is a brief summary of symptoms that assist the gardener to discover the causal agent(s).

Symptoms of disorders within the plant may be expressed as leaf yellowing (chlorosis), browning (necrosis), burning (white coloration due to loss of chlorophyll in leaves), deformation of leaves and growing tips, and stunting of overall growth. The first thing to observe with a nutrient disorder is the location of the affected tissue. Leaves will in general show the symptoms first. If it is a root problem due to disease or lack of oxygen, examination of the roots will reveal that they are not turgid and white, but, slimy and brown (Figure 10.1). The plant will wilt during high light periods as the water loss by transpiration is greater than the roots' ability to take up sufficient water.

The location on the plant of symptoms is the first clue as to the cause of the disorder. Focusing on leaf symptoms, if the lower leaves are expressing yellowing, browning, or spots first (Figure 10.2), then the group of nutrients responsible for the disorder would be those of "mobile" elements. Mobile elements can be retranslocated within the plant from the lower older tissue to the younger tissues in the top of the plant. These elements include N, P, K, Mg, Zn, and Mo. Initial symptoms will be a yellowing (chlorosis) followed by browning or drying (necrosis) of leaf tissue. If the

FIGURE 10.1 Healthy plant compared with a diseased one. (Drawing courtesy of George Barile, Accurate Art, Inc., Holbrook, New York.)

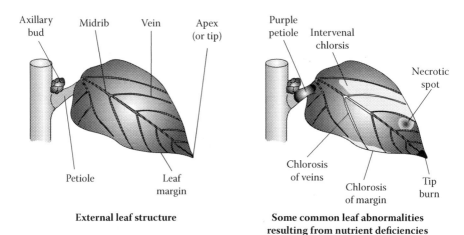

External leaf structure   **Some common leaf abnormalities resulting from nutrient deficiencies**

FIGURE 10.2 Common symptoms of nutrient disorders on leaves. (Drawing courtesy of George Barile, Accurate Art, Inc., Holbrook, New York.)

symptoms appear in the young leaves at the tip of the plant this disorder is a result of a lack of "immobile" elements that cannot move from the older plant parts to the growing tip. These immobile elements are Ca, B, Cu, Mn, S, and Fe. To determine which of these is the cause of the disorder there are "keys" composed of a dichotomous table allowing you to make a number of alternative choices (Figure 10.3). Each selection narrows the possible causes until in the final step there is a single element identified.

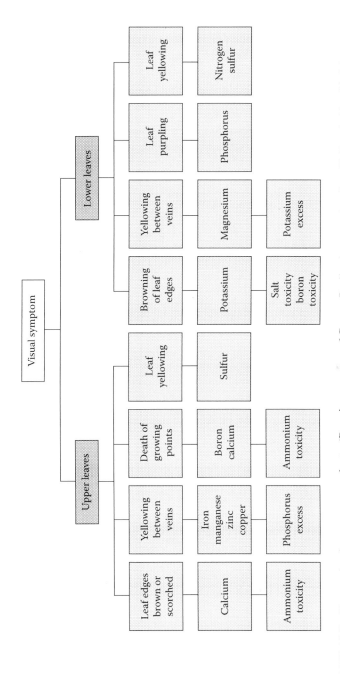

FIGURE 10.3 Key to visual symptoms on plants. (Drawing courtesy of George Barile, Accurate Art, Inc., Holbrook, New York.)

It is critical to recognize any symptoms occurring at an early stage of the plants' expression of these stress clues because as the disorder goes on without correction, the symptoms expand progressing from simple yellowing spots to complete yellowing and necrosis. At that stage, it is very difficult to know the first form of symptoms as they spread throughout the plant giving it an overall chlorosis, necrosis, and deformations of tissues. In addition, as the stress becomes more severe, it will be difficult, taking a lot of time to correct it once identified. The loss of the plant's health may become permanent or even result in its death. Yields will be greatly reduced as the stress is not corrected. The stress may begin as a cause from a single element and then as it progresses, other element uptake is slowed or blocked and the plant suffers from multiple disorders.

A very useful procedure when a symptom first appears is to immediately change the nutrient solution. That is, make up a new batch. At the same time, to determine the exact cause send a nutrient and/or tissue sample to a laboratory for analysis. Similar to soil analysis, the laboratory will give you guidelines as to what the normal levels of each nutrient should be in the solution or in the plant and direct you to make adjustments in the nutrient solution formulation.

MOBILE ELEMENTS

Here is a summary of deficiencies of mobile elements (first symptoms on older leaves) (Figure 10.3) and possible remedies.

NITROGEN

Lower leaves become yellowish green and growth is stunted.

Remedies

Add calcium nitrate or potassium nitrate to the nutrient solution.

PHOSPHOROUS

Stunted growth of plant, a purple color of the undersides of the leaves is very distinct and leaves fall off prematurely.

Remedies

Add monopotassium phosphate to the nutrient solution.

POTASSIUM

The leaflets on older leaves of tomatoes become scorched, curled margins, chlorosis between veins in the leaf tissue with small dry spots. Plant growth is restricted and stunted. Tomato fruits become blotchy and unevenly ripen.

Remedies

Apply a foliar spray of 2% potassium sulfate and add potassium sulfate to the nutrient solution.

MAGNESIUM

The older leaves have interveinal (between veins) chlorosis from the leaf margins inward, necrotic spots appear.

Remedies

Apply a foliar spray of 2% magnesium sulfate. Add magnesium sulfate to the nutrient solution.

Note: When applying foliar sprays, if in a greenhouse, avoid doing so during high sunlight conditions as that can cause burning of the leaves. Apply in the early morning while the sun and temperatures are low.

ZINC

Older and terminal leaves are abnormally small. The plant may get a "bushy" appearance due to the slowing of growth at the top.

Remedies

Use a foliar spray with 0.1%–0.5% solution of zinc sulfate. Add zinc sulfate to the nutrient solution.

IMMOBILE ELEMENTS

The following is a summary of deficiencies of immobile elements (first symptoms appear on the younger leaves at the top of the plant) (Figure 10.3) with suggested remedies.

CALCIUM

The upper leaves show marginal yellowing progressing to leaf tips, margins wither, and petioles curl and die back. The growing point stops growing and the smaller leaves turn purple-brown color on the margins, the leaflets remain tiny and deformed. Fruit of tomatoes show blossom-end rot (BER) (leathery appearance at blossom ends of the fruit).

Remedies

Apply a foliar spray of 1.0% calcium nitrate solution. Add calcium nitrate to the nutrient solution.

SULFUR

Upper leaves become stiff and curl down, leaves turn yellow. The stems, veins, and petioles turn purple and plant growth is restricted.

Remedies

Add potassium sulfate or other sulfate compound to the nutrient solution. A sulfur deficiency is usually rare because it is added to the nutrient solution by use of potassium, magnesium, and other sulfate salts.

IRON

The terminal leaves start turning yellow at the margins and progress through the entire leaf leading eventually to necrosis. Initially the smallest veins remain green giving a reticulate pattern. Flowers abort and fall off, growth is stunted and spindly in appearance.

Remedies

Apply a foliar spray with 0.02%–0.05% solution of iron chelate every 3–4 days. Add iron chelate to the nutrient solution.

BORON

The growing point withers and dies. Upper leaves curl inward and are deformed having interveinal mottling (blotchy pattern of yellowing). The upper smaller leaves become very brittle and break easily.

Remedies

Apply a foliar spray of 0.1%–0.25% borax solution. Add borax or boric acid to the nutrient solution.

COPPER

Young leaves remain small, margins turn into a tube toward the midribs in tomatoes, petioles bend downward, and growth is stunted to get a "bushy" appearance of the plant at the top.

Remedies

Use a foliar spray of 0.1%–0.2% solution of copper sulfate. Add copper sulfate to the nutrient solution.
Note: Whenever applying a foliar nutrient spray, apply it first to a few plants and wait to apply it to all plants for about a day to be sure that no burn occurs from the spray.

MANGANESE

Middle and younger leaves turn pale and develop a characteristic checkered pattern of green veins with yellowish interveinal areas. Later small necrotic spots form in the pale areas. Shoots will become stunted.

Remedies

Apply foliar spray of 0.1% manganese sulfate solution. Add manganese sulfate to the nutrient solution.

MOLYBDENUM

All leaves show a pale green to yellowish interveinal mottling, usually progresses from the older to the younger leaves.

Remedies

Apply a foliar spray of 0.07%–0.1% solution of ammonium or sodium molybdate. Add ammonium or sodium molybdate to the nutrient solution.

You will note that distinguishing among the symptom differences, especially with copper and molybdenum is difficult. The differences among iron, boron, and manganese are very prominent with the effects on the growing points and the distinct checkering coloration of manganese.

A deficiency in calcium is very similar to that of boron in the growing point; however, calcium will cause the BER on the fruit of especially tomatoes and peppers. Nonetheless, always remember to check the moisture level in the substrate and adjust irrigation cycles to give adequate watering so that wilting of the plant does not occur, as such water stress would be the first cause of the BER symptom.

PHYSIOLOGICAL DISORDERS

These disorders occur from environmental stresses such as high relative humidity, excessive temperatures (either high or low), very high light intensity, and incorrect irrigation. Often unfavorable environmental conditions cause upsets in nutrient uptake and therefore will also appear as a nutrient disorder. These disorders, including the entire nutrient disorders described in this chapter, are not encountered only in hydroponics, but are very common in soil growing also.

Blossom End Rot (Tomatoes, Peppers, Eggplants)

A brown, leathery tissue forms at the blossom end of the fruit (Figure 10.4).

Causes and Remedies

Calcium deficiency, water stress due to insufficient irrigation frequency or too much on compact clay soils that causes poor root aeration. In hydroponics, it is a lack of irrigation cycles, especially under high temperatures and light intensity. It is often a calcium deficiency induced by poor irrigation practices. Adding calcium will not rectify the problem if the irrigation frequency is not corrected.

Fruit Cracking (Tomatoes, Peppers, Eggplants)

Cracks radiate from the stem end, especially on maturing fruit (Figure 10.5).

Causes and Remedies

Poor irrigation is the cause of water deficit, especially under high temperatures, when an irrigation cycle is initiated the water is taken up very rapidly by the plant that directs it to the fruit where the sudden expansion is too fast for the skin to expand and it cracks. This can be prevented by avoiding high temperatures with shading and maintaining uniform soil moisture levels. Start irrigation cycles 1–2 hours after sunrise and the last one no later than 1 hour before sunset.

FIGURE 10.4   Blossom-end rot of tomato fruit.

Blotchy Ripening (Tomatoes)

Fruit color is uneven with brown vascular tissue inside the fruit.

Causes and Remedies

There are a number of environmental and possible induced nutritional disorders. Low light intensity, cool temperatures, high medium moisture levels, high nitrogen and low potassium are all potential causes. Avoid this condition by adding

FIGURE 10.5   Cracking of tomato fruit.

supplementary lighting, or using less irrigation cycles under low light conditions and lower nitrogen levels in the nutrient solution.

GREEN SHOULDER, SUNSCALD (TOMATOES, PEPPERS, EGGPLANTS)

The top shoulder area of the fruit remains a blotchy green while the rest of the fruit is colored. This is particularly common in some tomato varieties. A lot of varieties have resistance to this disorder. With peppers and eggplants, a blackened leathery spot appears on the fruit as a result of sunscald.

Causes and Remedies

The cause is high temperatures combined with direct sunlight striking the fruit. This can be prevented by keeping good leaf growth above the ripening fruit and in a greenhouse provide shading in the hot, summer months. Also, if peppers have a lot of fruit developing at a given time, this high production slows the plant growth and fruit forms near the tops of the plants where few, or small developing leaves cannot shade the fruit sufficiently. You can add nitrogen to the nutrient solution to promote more vegetative growth in the plant. In addition, with peppers, do not permit more than five to six fruits to form on each stem of the plant. Thin, if necessary, to reduce the fruit load and this will also give you larger fruit.

CATFACING, MISSHAPEN FRUIT (TOMATOES, PEPPERS, EGGPLANTS)

This is fruit distortion with protuberances and indentations (Figure 10.6).

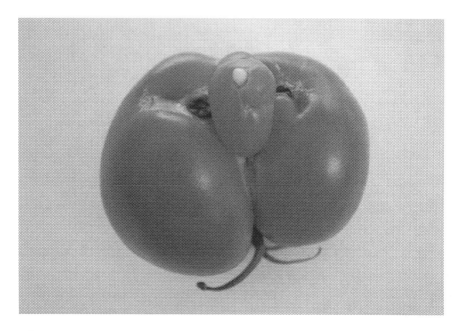

FIGURE 10.6   Catfacing of tomato fruit.

Causes and Remedies

High relative humidity and low light levels cause poor pollination. Avoid these environmental conditions with ventilation and addition of supplementary lights, especially with indoor growing to increase light intensity.

CROOKING (CUCUMBERS)

This is the bending of the fruit as it expands (Figure 10.7).

Causes and Remedies

The causes include any poor temperatures that cause slowing of growth, fruit hanging up on leaves, tendrils (long stringy appendages of the cucumber) attaching to the fruit, mechanical damage, or pest injury during the rapid fruit expansion. The cure is to keep good temperatures and avoid the other causes by proper training of the plant as described in Chapter 24.

ABORTION OF FRUIT (CUCUMBERS, PEPPERS, EGGPLANTS)

The fruit gets soft, yellows, and shrivels when very small.

Causes and Remedies

This condition can be caused by poor nutrition, too heavy of fruit load, low light, and improper training of the plant. Keep the plant pruned and thin the fruit set if necessary.
    Note: Proper training of the plants is presented in detail in Section VI, Chapter 24.

FIGURE 10.7   Crooking of cucumber fruit.

SUMMARY

In summary, it is essential for good production to recognize any nutritional and/or environmental-induced disorders in plants early from symptoms they express when under stress. Their early correction will prevent losses in yields and their decline in vigor. There are many books available with photos describing these symptoms and keys to assist in identifying nutritional disorders such as in my book, *Hydroponic Food Production*.

Section IV

Hydroponic Systems

11 Substrates to Use and Their Sources for Hydroponics

Plants will grow in most media as long as they get water, oxygen, and nutrients. Of course, not all are well suited to provide optimum growth. Heavy clay soils are cold and have so much water that oxygen to plant roots is restricted limiting growth. There are a number of qualities that must be considered when choosing a substrate for hydroponic growing. Here are some important characteristics.

Structure: The structure must be durable for at least one or more crops and not break down into small particles that will impair oxygenation to plant roots.

Composition: The particles of the substrate must not react with the nutrient solution or release elements into the water as that will upset the balance of the nutrient solution. For example, calcareous rock releases calcium and magnesium causing the pH to rise above optimal levels. Some substrates such as coco coir must be well washed to remove any residual sodium chloride as often the coconut husks are found near salt water where the palm trees are growing. Bark and sawdust must be sourced from Douglas Fir, Hemlock, or Redwood timber as they are known not to release any turpines or other resins found naturally in pine and cedar wood. Rice hulls have to be aged for some time in piles that are watered, to permit those with embryos to germinate. Composting will generate heat and kill the seedlings of the rice. Alternatively, the rice hulls can be burned under a smoldering fire to carbonize them. This process gives the rice hulls less smooth surface so that it will retain more water than raw husks. The burning process will also sterilize the rice hulls and kill any embryos present.

Sterility: Substrates for hydroponic culture must be free of pest and disease organisms. If you are uncertain as to the sterility of the medium, heat it to 160°F (71°C) for half an hour to kill the organisms. This can be done in a kitchen stove oven or if there is intense sunlight and high temperatures, place the substrate on a black polyethylene and also cover it with a black polyethylene. This process should be okay after about 1 week. Gravel and some coarse sands could be sterilized using a 10% bleach solution. Other finer substrates like perlite can be pre-treated with "Zerotol," a hydrogen dioxide compound that is highly oxidizing. It must be soaked 2–3 days prior to sowing or transplanting.

Water Retention: The hydroponic substrate must not have properties of very high or very low water retention. However, the acceptable water retention will also be a function of the type of hydroponic system. Coarse gravel can be used with a sub-irrigation or ebb-and-flow watering system. The water (nutrient solution) will enter the void spaces among the rock particles when the bed is being flooded and wets the particle surfaces. As the solution drains back out it will pull air into the substrate. This method of hydroponics functions well with coarse substrates.

With fine substrates like rockwool, sawdust, coco coir, and peatlite mixes, use a drip irrigation system that applies the solution on the top and spreads through the substrate by capillary action and drains out the bottom of the container.

Water retention must not be excessive causing lack of oxygenation or be insufficient to cause the substrate to dry quickly and starve the plant of both water and nutrients.

Root Support: The substrate must allow roots to easily penetrate between the particles and anchor the plant as the roots enter the void spaces seeking water. If the substrate is too fine, the roots cannot spread readily into the medium, whereas, if it is too coarse, the plant roots will not be able to hold onto the particles and the plant can easily fall over. Most long-term crops need a substrate to anchor their roots and take up oxygen, water, and nutrients.

Availability and Cost: For small hydroponic gardens, the cost is not an important factor as relatively small amounts of substrate are needed. Many types of substrates are available for hydroponics and may be purchased online or at hydroponic shops.

MEDIA (SUBSTRATES)

The following is a description of many of these suitable substrates for hydroponic growing. When we discuss the various hydroponic systems for different plants, specific substrates are recommended.

1. **Gravel**: This was one of the original substrates used in hydroponic culture from the 1940s through the 1960s. It was the substrate introduced by Dr. W.F. Gericke to establish outdoor hydroponic operations during World War II on non-arable islands of the Pacific. In many of those islands, volcanic rock was suitable for hydroponics and it was plentiful.
   Choosing the most desirable rock takes into account a number of characteristics or properties. It should be irregular in shape (crushed is best), free of fine particles (fine sand to silt), and should be aggregate of ½–¾″ in diameter (Figure 11.1). The particles must be of igneous (volcanic) origin and be structurally stable. They must retain adequate moisture in void spaces yet drain well to provide oxygen to the plant roots. Avoid the use of calcareous gravel (sedimentary origin-like limestone) as their release of calcium carbonate will continually alter the pH.
2. **Pebbles (Bird's Eye and Pea Gravel)**: These gravels have round, smooth surfaces with size ranging from a ⅛″ to ¼″ diameter (bird's eye vs. pea gravel) (Figure 11.1). Since their surface is less irregular, the smaller size is critical to their suitability. If you select gravel larger than pea gravel size, it must be with a sub-irrigation system as described earlier for gravel. Again, this material must be of igneous origin. With this substrate, I have used drip irrigation or soaker hoses to distribute the nutrient solution to the base of the plants. However, due to the rapid percolation, it is best to use a recycle system of hydroponic culture and more frequent irrigation cycles than for

other finer materials. These media, like gravel, can be sterilized with a 10% bleach solution between crops. Always remember to flush the substrate with raw water after using a bleach solution to remove any residual chlorine.

3. **Leca (Expanded Clay)**: This fired clay (Figure 11.1) is sometimes called "Haydite" or "Herculite" as it is light weight and is used in construction. It has good water retention from its irregular surface and is especially suitable for small hobby units and indoor gardening. It does fracture with time resulting in the release of fine sand and silt, but with a hobby unit it can be replaced when this build-up of fines occurs.

4. **Scoria**: Crushed rock from volcanic origin is highly vesicular (full of holes from escaping gases during cooling from molten magma) (Figure 11.1). It is usually dark brown, black, or purplish red and light in weight. It can also be called cinder. It is fairly good in retaining water and at the same time gives good oxygenation. Various particle sizes can be used for hydroponics, ranging from ½″ to less than ⅛″ in diameter.

5. **Sand**: The best sand is river sand of igneous origin. It must be well washed by the quarry operator. This washed river sand is available from most aggregate suppliers. Do not use mortar sand as its fine particles cause puddling, water coming to the surface when vibrated. The settling of the very fine particles will reduce any void spaces and eliminate the available oxygen. If properly screened to eliminate any particles of diameter over 2 mm (0.0625″) and under 0.6 mm or 0.025″, the sand will drain freely and provide adequate oxygen to plant roots.

6. **Sawdust**: Where large forest industries are present, sawdust is a good medium provided the source is from Douglas fir or western hemlock trees.

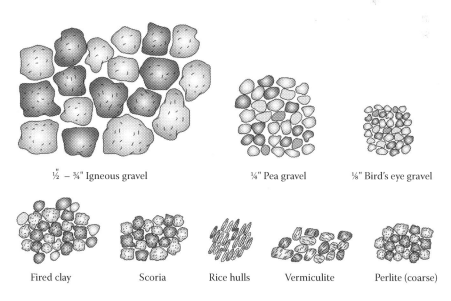

½″ – ¾″ Igneous gravel   ¼″ Pea gravel   ⅛″ Bird's eye gravel

Fired clay   Scoria   Rice hulls   Vermiculite   Perlite (coarse)

FIGURE 11.1 Various substrates. (Drawing courtesy of George Barile, Accurate Art, Inc., Holbrook, New York.)

Western red cedar and pines should not be used as their resins are toxic to plants. In many locations, logs are floated in barges on the ocean and they collect sodium chloride from the water. Test the sawdust for sodium chloride content and leach it thoroughly with pure raw water before planting.

7. **Peat**: Peat is partially decomposed freshwater marsh or swamp vegetation. Use peat from sphagnum moss, as some of the other types are very fine and hold too much water. Other peats from sedges, reeds, and hypnum moss decompose rapidly upsetting the structural integrity of the medium. Peat is readily available in 3.8 cubic foot bales in compressed form. There are many blends available mixed with various percentages of perlite, vermiculite, or Styrofoam particles to add aeration to the medium. It also comes with the beneficial microorganism, Mycorrhizae. For example, Premier Tech Horticulture offers a line of peat-based substrates. Their "Pro-Mix BX Mycorrhizae" is a general purpose peat-based medium designed for greenhouse/indoor growing and contains a mycorrhizal inoculum. The symbiotic fungi colonize the root systems of plants and increase water and nutrient acquisition. The ingredients of this mix include perlite, vermiculite, dolomite lime to adjust the pH, and a wetting agent to assist in absorption of water during its initial dry state.

8. **Peatlite Mixes**: There are various mixtures of peat-based media that have been developed with extensive research by various universities. The two most popular ones are the UC mix (University of California) and the Cornell "Peat-Lite" mixes (Cornell University, New York). These mixtures are combinations of peat, sand, perlite, pumice, and vermiculite with added nutrients and dolomite lime to adjust the pH. The pH of peat is very low so a base, such as dolomite lime, is needed to increase the pH. You may refer to these and other mixes on the Internet or in my book *Hydroponic Food Production*. It is easiest to simply purchase a complete mix such as the "Pro-Mix."

9. **Redwood Bark**: As the name suggests this is bark from redwood trees. It is much coarser than sawdust with particle diameters from ⅛″ to ⅓″ or larger. This is usually the preferred medium for the growing of orchids, but not commonly used in other hydroponic cultures.

10. **Rice Hulls**: This is the outer husk or shell of rice (Figure 11.1). This is a by-product of rice milling and is a waste product. They will last from 3 to 5 years without decomposing. They have a smooth surface so do not retain water readily and have poor capillary (lateral) movement of water. If they are burned by a smoldering fire, the surface is improved to give more water retention. Rice hulls are best mixed with peat or coco coir, usually at 20% rice hulls.

11. **Vermiculite**: Vermiculite is expanded mica through a heating process to form spongy particles (Figure 11.1). Heating also insures that the medium is sterile. With the layers (cleavage) formed in each particle water retention is high. The irregular shape of the particles also creates ample void spaces to retain moisture and make oxygen available to the plants. The horticultural vermiculite comes in four grades from the coarse material

of particle size from 0.2″ to 0.3″ in diameter to the fine material with particle size of 0.04″ (1 mm). Use the coarse grade for hydroponic growing and the fine to medium grade for seed germination. Be careful not to press the vermiculite together when adding water as that will destroy its structure.

12. **Perlite**: This comes from volcanic pumice. It is crushed and heated to expand the particles (Figure 11.1). It is very light-weight and is sterile due to the process of heating. A coarse particle size between 2 mm and 3 mm (0.065–0.13″) is best for hydroponics. Particles are irregular, but more structurally stable than that of vermiculite. It contains no nutrients and has a pH of 6.0–8.0. A fine grade is used for seed germination while the coarse grade is best to use in plant growth. It may be used by itself or in mixtures with peat as explained earlier with peatlite mixtures.

13. **Coco Coir**: Coco coir is ground-up dried coconut palm husk. The processed material includes coir fibers and pulp. Most comes from Indonesia, India, Sri Lanka, Thailand, and Brazil. This substrate is now becoming the principal one used in large greenhouse operations basing their production on "sustainable" agricultural technology. The reason for that is that coconut husks are a renewable product.

A number of different forms of coco coir are available in the market. There is a compressed bale or block that expands upon the addition of water. The blocks will expand to double their size within 15 min of soaking in water. A product created especially for greenhouse hydroponics is the slab form where the coco coir is encased in a long plastic bag (Figure 11.2). It measures about 1 m (39.4″) long by 6–8″ wide by 3–4″ thick. These slabs are aimed at growing vine crops such as tomatoes, cucumbers, peppers, and eggplants. They are available in different grades of coco coir with varying percentages of the coir pulp and fibers. This imparts a series of air-holding capacities from 20% to 40% air capacity at saturation. The choice depends upon the crop to be grown. Slabs of half husk chips and half coco pith are recommended for tomatoes and peppers while those of 100% husk chips are for cucumbers. Coco coir plugs and blocks are also available to start and transplant seedlings.

Coco coir can be mixed with perlite, rice hulls, or vermiculite similar to the peatlite mixtures described earlier. One fact to consider is the source of the coco coir as that from coastal areas may contain sodium chloride. In that case, it is necessary to flush the coco coir with pure raw water to remove the sodium chloride. Most manufacturers do that during processing so their products will not require initial flushing.

14. **Rockwool**: Rockwool is made from basaltic rock (solidified lava) that is liquefied at 1500°C (2732°F). It is spun and then pressed into sheets that are cut into slabs, blocks, and cubes (Figure 11.3). The slabs have similar measurements to coco coir slabs. Rockwool is slightly alkaline but inert and does not decompose. With 95% pore spaces, it has good water-holding capacity. The pH between 7 and 8.5 must be adjusted before seeding or

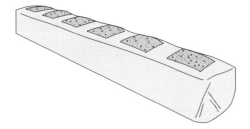

FIGURE 11.2 Slab of coco coir. (Drawing courtesy of George Barile, Accurate Art, Inc., Holbrook, New York.)

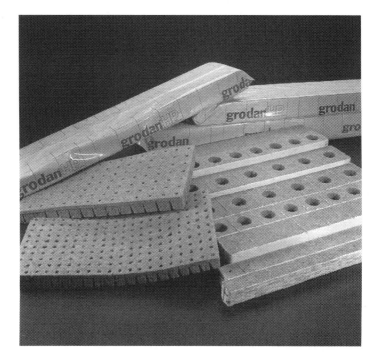

FIGURE 11.3 Rockwool cubes, blocks, and slabs. (Courtesy of Botanicare, Tempe, AZ.)

planting by saturation with an acid nutrient solution to reach an optimum pH of 6.0–6.5 for most vine crops. A drip irrigation system is used as is the case with coco coir, peatlite, perlite, and other fine mixtures. Managing the irrigation cycles is crucial to success as rockwool needs a 20%–30% leachate (drainage) to prevent any mineral build-up in the substrate. Emphasis now is on recycling the leachate back to a tank where its pH and electrical conductivity are adjusted with an injection system before its reuse in the crop. Special trays are now on the market to collect the leachate from the slabs and return it to a central batch adjustment tank. The same principle of recirculation of spent nutrient solution with rockwool applies to home gardening. These details are described in Section V.

Rockwool cubes and blocks are the standard method of starting your plants for hydroponic culture. Even if you do not use rockwool as a final growing medium, it is still best to start your seedlings in rockwool cubes that are transplanted to larger rockwool blocks before their final transplanting to the growing area. Rockwool cubes are available in $1\frac{1}{2}'' \times 1\frac{1}{2}'' \times 1\frac{1}{2}''$ size for tomatoes, cucumbers, peppers, and eggplants, and $1'' \times 1'' \times 1\frac{1}{2}''$ deep smaller ones for lettuce and herbs. The blocks come in several sizes as well, with larger ones for plants that you wish to keep in the seedling area longer. They are available as $3'' \times 3'' \times 2.5''$, $3'' \times 3'' \times 4''$, $4'' \times 4'' \times 2.5''$, and $4'' \times 4'' \times 3''$. The blocks have round holes of $1\frac{1}{2}''$ diameter by $1\frac{1}{2}''$ deep to fit the cubes during transplanting.

Loose rockwool granules can be used in containers and/or mixed with other media of larger aggregates such as scoria to improve moisture retention.

Do not confuse rockwool with insulating material, such as the pink or sandy color bats for your home, since it is not suitable for growing plants. It absorbs a lot of water and remains saturated excluding oxygen from plant roots. It also collapses in on itself structurally when squeezed.

15. **Water**: The use of water alone is the choice of medium for growing lettuce, basil, arugula, and some herbs. The systems discussed later are those of the raft culture and nutrient film technique (NFT). In these systems, the plant roots are suspended directly into the nutrient solution. A supporting cover of Styrofoam boards (raft culture) and channels or gutters in the case of NFT keep the plants above the nutrient solution (Figure 11.4).

Water, of course, is the medium for the essential elements to form the nutrient solution. Water must be very pure with few or no extraneous elements. It must not have sodium chloride present in excess of 50 ppm. If your source of water is from a city reservoir, it will be safe to use in hydroponics. If it comes from well water, you will need to have it tested to determine what elements are present and at what levels. Once you know the levels of the elements, you can determine if it is necessary to adjust the nutrient formulation to take these into account. You simply subtract the concentration of each of the elements present from the total you wish to add and calculate the balance of each you must add to get your optimum formulation. Often raw water is "hard." That means that there is calcium and magnesium carbonates present, so the pH will have to be lowered with an acid. Remember that both calcium and magnesium are essential macro-elements; therefore, you will adjust the levels to add by those levels present in the raw water. In effect, this will save on the amount of calcium and magnesium used in the nutrient solution. Be particularly aware of any micronutrients that are present in raw water, especially boron as sometimes there is sufficient in the raw water that it will not have to be added to the formulation.

FIGURE 11.4 Plant roots suspended in a nutrient solution of raft culture. (Courtesy of CuisinArt Golf Resort and Spa, Anguilla.)

If the water is just hard, you can purchase a prepared nutrient formulation specifically for hard waters assuming that you may not wish to make up your own formulation.

Water analyses are offered by a number of laboratories that also do soil and tissue analyses. Please refer to the reference section for specific websites. Whether growing in soil or hydroponics, it is important to know the elements and their concentrations in the raw water in order to adjust your fertilization program to avoid excess levels of any element. With hydroponics, it is much easier to make these adjustments to your nutritional program compared to soil growing.

12 Small Indoor Hydroponic Units
Types and Construction

INTRODUCTION

There is a great variety of hobby indoor hydroponic systems in the market. All hydroponic systems, whether hobby or commercial size, can be placed within two groups according to their circulation of the nutrient solution. One is the recirculated or closed system; the other is that which is a nonrecycled or open system. Clearly, indoors in your home you must use a recirculation or a waste water collection system to prevent flooding your basement or other location having the hydroponic facility. In terms of nutrient management, recycled systems are more challenging than open systems. This is due to shifts in pH and nutrient levels of each element within the nutrient solution. Plants do not take up all elements at the same rate or quantity, especially under varying environmental conditions. As a result, each essential element in the nutrient solution changes with time at differing rates. This causes shifts in pH and the total salts in the solution measured by electrical conductivity (EC). However, in small household units when you find large changes in EC occurring you can simply dump the old solution (use it on your houseplants) and make up a new batch. The pH can easily be adjusted at any time with an acid or base as described earlier in Chapters 7 and 9.

Open systems are easier to manage since each irrigation cycle adds new solution. You must add sufficient solution to get some leachate as was discussed in Chapter 6. To overcome problems of disposal of the leachate (drainage) during each irrigation, simply place the growing units above a collection pipe that can conduct the spent solution to a waste reservoir. The waste solution can be used to water your house plants or even outside soil garden. Open systems have the advantage that if a disease got into the system, it would not spread back into the plants as it may in a recycled system.

SMALL INDOOR SYSTEMS

The following discussion of small indoor units is organized from the most simplified ones to more complex ones. These are all recycled systems. In Chapter 13, larger systems of both recycled and nonrecycled systems are presented.

NURSERY TRAY

This is a very simple system using a standard plastic flat (10½″ × 21″) without holes and a compact cell tray insert (Figure 12.1). These compact cell trays come in many different sized cells or cube partitions. They are readily available in garden centers. You may even have some left over from the last time you bought bedding plants for your outside garden. The best size for growing herbs and small lettuces is either a "24" or "36" cell tray. Cut out one corner of sufficient size to fit a plastic 1-gallon container. The plastic container must have a large lid of about 4″–5″ in diameter. Drill a ¼″ diameter hole in the middle of the lid and glue a ¼″ wide cork ring of 3″ in diameter. If you cannot find such cork rings, you may purchase a sheet of ¼″ thick cork and cut one from it. Remove a small gap from the ring, about ½″, so that it is not complete (Figure 12.2). This gap is going to allow the solution to flow from the bottle to the bottom of the tray. Place your finger on the hole in the lid after filling the bottle with nutrient solution and invert it into the corner of the flat where the corner of the cell tray was removed. Solution will flow from it until the level in the tray reaches the level of the hole in the lid. From then on, it will automatically siphon whenever the solution level in the tray falls below the lid face allowing air to enter the bottle. Refill the solution bottle as it empties from the plants' water demand.

Use either perlite or vermiculite or a mixture of 20% peat with 80% of either perlite or vermiculite as the substrate. Place the substrate in the cell tray; water it with a watering wand or watering can to thoroughly moisten the medium prior to sowing the seeds. You must use raw water only until germination of the seeds takes place and

FIGURE 12.1 Simple hydroponic nursery tray system. (Drawing courtesy of George Barile, Accurate Art, Inc., Holbrook, New York.)

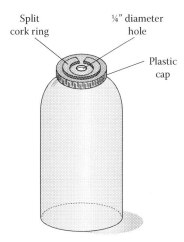

Split cork ring

¼" diameter hole

Plastic cap

FIGURE 12.2 Nutrient reservoir bottle with cut ring and hole. (Drawing courtesy of George Barile, Accurate Art, Inc., Holbrook, New York.)

the seedlings form their first true leaves, then use a half-strength nutrient solution for the next few weeks before using full strength solution. Do not use the inverted bottle reservoir in the tray until the seedlings are ready for the nutrient solution. You must harvest the herbs, basil, arugula, and lettuce fairly small otherwise they will extend due to the tight spacing restricting light to each individual plant. These types of crops would normally be for 4–6 weeks depending upon their nature of growth. The tray may be used to grow mesclun mixes of baby lettuce, herbs, arugula, beets, mustards, mizuna, chard, and spinach.

WICK SYSTEM

The wick system is an old system, but it works, especially for individual pots. This is a very simple form of hydroponics. It can be set up as a single pot or a series of pots or a tray of medium sitting on top of a nutrient reservoir (Figure 12.3). One or more wicks are positioned in the substrate and hang down into a reservoir of nutrient solution below. Capillary action moves the solution from the reservoir to the base of the plants as the medium dries. Use cotton or nylon fibrous rope. Bury one end of the wicks in the substrate close to where the plants are growing and let the other end dangle down into the nutrient reservoir below. It is best to flare the ends of the wicks to get better uptake and distribution of the solution. The choice of medium to use in this system is a 50/50 mixture of perlite and vermiculite.

MANUAL SYSTEMS

These are any simple systems whereby no electricity is needed for pumps or other components. It is simply the addition of nutrient solution to the growing unit by raising and lowering of a tank during an irrigation cycle (Figure 12.4). This can be done by hand. When the reservoir is raised, the solution will flow from the tank

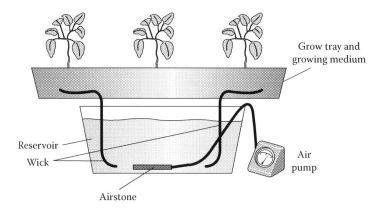

FIGURE 12.3 Wick system with wicks suspended to solution tank below. (Drawing courtesy of George Barile, Accurate Art, Inc., Holbrook, New York.)

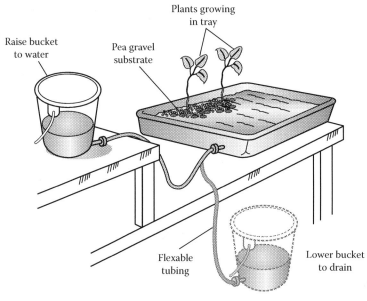

Manual Bucket System

FIGURE 12.4 Manual system of raising reservoir to irrigate plant tray and lowering it for drainage. (Drawing courtesy of George Barile, Accurate Art, Inc., Holbrook, New York.)

to the plant tray and when the reservoir is lowered below the level of the growing bed the water will drain back into the tank. This technique is a flood and drain system. The frequency of irrigation cycles depends upon the nature of the growing substrate. You may use perlite alone, a mixture of perlite and peat (50/50), perlite and vermiculite (50/50), peatlite mixture, coco coir mixture, gravel, pebbles, coarse sand, scoria, or expanded clay. The finer the material, the less irrigation

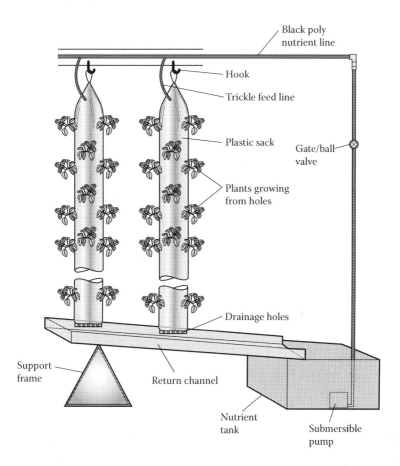

FIGURE 12.5 Small sack culture system design. (Drawing courtesy of George Barile, Accurate Art, Inc., Holbrook, New York.)

cycles needed per day. With peatlite or coco coir alone, one irrigation per day would be sufficient.

This same principle may be applied to growing herbs, lettuce, arugula, basil, and even flowers in vertical sacks (Figure 12.5). Sacks are constructed of 6-mil layflat polyethylene that will give a diameter of 6–8″ when filled with a substrate. Cut the sacks about 6 ft long. Tie the bottom with string and fill the sack with moistened peatlite or coco coir. To make it lighter make a mixture of 70% coco coir and 30% rice hulls.

This medium must be moistened prior to placing it in the sack as water will not wet the dry substrate thoroughly when placed in the sack. Break up the peat or coco coir and mix it with some water in a wheelbarrow. Add a little water at a time and mix it uniformly. Test the moisture level in the medium by taking a handful of it and squeeze it until you see a small amount of water coming out. When you release pressure on it, the ball of medium should slightly break apart, but not collapse. At that point you have adequate water incorporated into the substrate. Be careful not to add too much water. That would be indicated by the ball not breaking slightly and excess moisture draining through your fingers upon squeezing it.

Once you fill the sack, tie the top of the sack leaving about 10″–12″ empty. Fold this end over and tie it again back to the sack. This will allow a loop by which you can support the sack to a frame above with a hook or rope. Use a 2 L (1/2 gallon) plastic water bottle as the solution reservoir. Cut a hole in the top of the sack immediately below the tie of the string. Insert the neck of the bottle in the sack through the hole. For a collection tank use a pail or plastic container with a top. Make some holes in the top of the lid and place it directly under the sack drain end. To irrigate take the water from the collection tank and fill the top irrigation bottle letting it percolate through the sack. Replace the collection pail under the sack to reuse the solution. Of course, as the plants use the solution you will need to add more new solution to the fill bottle.

Cut 1″ holes in the sack going down the sack spacing them at 6–8″ depending upon the crop you wish to grow. Use closer spacing for smaller plants. Stagger the position of the holes to improve light penetration to the crop. Start seedlings in rockwool or Oasis cubes and transplant to the sacks once they are about 2″ tall (usually 3–4 weeks old).

You can do this same vertical culture using 6″ diameter polyvinyl chloride (PVC) pipe instead of the plastic sacks (Figure 12.6). Cut 1″ holes in the pipe for the transplants similar to that for the sacks. However, it is best to use 1″ 90° elbows for the plant sites as shown in Figure 12.6. This will support the plants during transplanting. The PVC pipe can sit on a collection pipe of similar diameter that would conduct the solution to a tank. Cut the collection pipe in half lengthwise and fill it with gravel substrate to permit unobstructed drainage to the collection tank. The vertical pipe can also be supported by the collection pipe with additional support to secure its position at the top. Place a one-gallon plastic bottle on top of the pipe.

Remove the bottle cap from the container and invert it into the top of the growing pipe. You can use this fill bottle as a funnel or simply keep it as a storage bottle with nutrient solution that you collect from the drainage of the vertical grow pipe.

In the preceding systems of sack and column culture, a substrate is used to grow the plants.

These sack and column culture systems can be fully automated by having a nutrient reservoir underneath with a submersible pump operated by a time clock. Such automated systems with their plumbing are shown in Figures 12.5 and 12.6. The column culture system may also be set up as an aeroponic method without any substrate. In that case, the nutrient line will have mist jets every 6″ along the pipe opposite the plant sites. The pipe may be installed inside the column or outside entering the column from above as shown in Figure 12.6 as alternative piping. If the column is used with a substrate, the outside plumbing method would be used with drip lines at the top of the column similar to the sack culture system. The plant sites should be arranged spirally down the column at 6–8″ centers.

Both sack culture and column culture are suitable only for low-profile plants like lettuce, arugula, basil, and herbs.

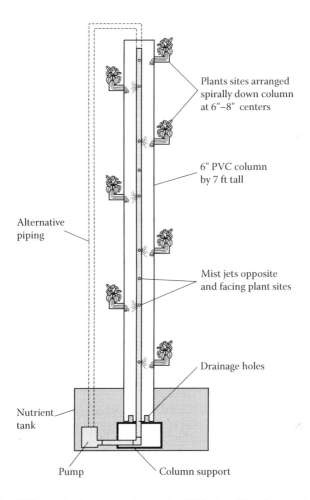

FIGURE 12.6 Self-contained column culture using PVC pipe. (Drawing courtesy of George Barile, Accurate Art, Inc., Holbrook, New York.)

WATER CULTURE (FLOATING OR RAFT)

If you wish to grow lettuce, arugula, basil, and some herbs, a simple raft culture system may be constructed. With this system construct a bed from concrete blocks or treated wood. The blocks can be 4″ thick by 8″ by 16″. Place them on edge to get an 8″ depth of the bed. Make the inside dimensions several inches wider on each side of a 4 ft × 4 ft frame. Line it with 12-mil polyethylene (black) or a 20-mil thick vinyl swimming pool liner. Fold the inner corners up like an envelope and glue these laps, if vinyl, using vinyl cement. If using polyethylene bring the liner onto the top of the cement blocks and hold it down with a perimeter piece of lumber. If you use 2″ × 8″ treated lumber, you can staple the liner to the top of the lumber and then nail a lathe around the perimeter to secure the liner edge uniformly.

Purchase a 4 ft × 8 ft × 1″ thick Styrofoam. Use the pink or blue "Roofmate" denser material as it will not break as easily as the less dense white material. Cut the board into 4 ft × 4 ft. Using a saw hole drill cut holes of ¾–⅞″ diameter so that the grow cubes (rockwool or Oasis) will fit snugly into the holes during transplanting. The holes are spaced 6″ × 6″ center to center. Make up nutrient solution to fill the raft bed to within 1″ from the top. The volume of water in this size of bed is as follows:

Volume = length × width × height
V = 4 ft × 4 ft × 7″/12″ = 9.33 cubic feet
Conversion to U.S. gallons: 1 cubic ft = 7.48 gallons
Therefore: 9.33 × 7.48 = 70 gallons

Start your seedlings as explained in Chapter 14.

Upon transplanting the seedlings to the raft system aerate the solution several times a day by beating the solution with a whisk. Alternatively, of course, you could add an air pump with a tube attached to an airstone in the bottom of the bed.

There are two main water culture systems, raft or floating and nutrient film technique. In this section the raft system is presented. This system is ideal for lettuce, arugula, basil, mint, watercress, and a few other herbs. It is not suitable for vine crops or long-term plants due to eventual lack of oxygenation to plant roots. A small indoor unit can be constructed of the following components.

Supplies

1. A plastic storage bin with lid for keeping clothes or other items is great for a nutrient reservoir (Figure 12.7). Use a large one at least 14″ wide by 18″ long by 8″ deep, between 12 and 15 gallons. It must have a relatively flat lid and the bin should be of an opaque color such as black or dark blue to prevent light from entering. If light comes in contact with the nutrient solution, algae will grow in it causing plugging of lines and unwanted build-up of slime in the system. Also, you can wrap the bin with aluminum foil to

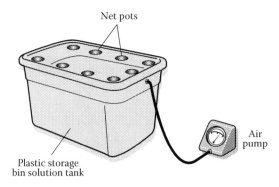

FIGURE 12.7   Simple deep flow system using a plastic storage bin with lid. (Drawing courtesy of George Barile, Accurate Art, Inc., Holbrook, New York.)

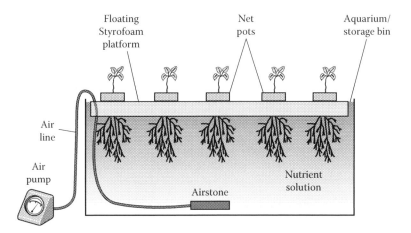

FIGURE 12.8 Simple raft system using a plastic storage bin, Styrofoam board cover, air pump, tubing, and so on. (Drawing courtesy of George Barile, Accurate Art, Inc., Holbrook, New York.)

prevent light and heat accumulation. The plants are seeded in rockwool or Oasis cubes and within 3 weeks transplanted to 2″ diameter net pots that support the plants in the lid of the storage bin (nutrient tank).

Alternatively, use storage or tote bins without a lid and cut a 1″ thick Styrofoam board to float on top of the solution (Figure 12.8). Prepare it as described earlier in "Manual Systems." Be sure to use the dense "Roofmate" type of Styrofoam. Cut ¾″ holes spaced 6″ × 6″ for lettuce, arugula, and basil. For smaller herbs make the spacing 4″ × 4″. The use of the Styrofoam cover over a plastic lid is that it insulates the solution below, but that should not be an issue if this is for growing plants inside your home under artificial lights.

2. An air pump, such as a fish aquarium pump with plastic tubing.
3. A fish aquarium airstone attached to the other end of the tubing. This is located in the nutrient reservoir to add oxygen to the solution. You may purchase this equipment at a pet store where fish and aquariums are sold. Use an airstone 4–6″ long to give lots of oxygen to the nutrient solution. Assembly is shown in Figures 12.7 and 12.8.

WATER CULTURE-NUTRIENT FILM TECHNIQUE (NFT) SYSTEM

The principle of the NFT system is to keep a constant flow of thin layer of solution through the plant roots in a channel or grow tray. This is a recirculation method returning the solution from the grow tray back to the nutrient reservoir below. The components are as follows.

SUPPLIES

1. One or two grow trays or gutters constructed from a 2″ PVC pipe. There are commercial NFT channels available at hydroponic stores. If you want

to use a 2″ PVC pipe, you will need the following fittings for each channel: one end cap, one 2″ 90° elbow with a 2″ × 1″ reduced bushing, 4″ of 1″ PVC pipe as the drain spout back to the reservoir underneath. If an elbow is used one end cap will be sufficient. However, it would be better to use one 2″ × 1″ reduced tee just before the end of the pipe and another end cap. That would be more effective than using the elbow and reduced bushing at the drain end of the pipe. Then, the end of the tee with the added 1″ spout would enter the lid of the reservoir.

2. Purchase a small submersible pump, such as a "Little Giant" fountain pump or a smaller submersible pump, at a hydroponic or irrigation store. Poly tubing to connect the pump with the inlet end to the growing channel. Drill a small hole on the top of the pipe at the front end where the solution will enter.

3. A nutrient reservoir of a storage bin as described earlier in "Floating System."

4. An aquarium air pump and airstone as was also described in "Floating System."

5. Small 2″ × 2″ round plastic mesh pots available at a hydroponic shop. These are needed to support the growing cube with the transplant. During transplanting place the seedling in the mesh pot, which sits in the hole in the growing channel.

ASSEMBLY

1. Drill one small hole in the storage bin lid near one end and the other at the opposite end. These must be just large enough to fit the poly hoses from the pump in the nutrient tank going to the NFT channels and the other from the air pump to the reservoir connecting the airstone.

2. The NFT channels will need 2″ diameter holes drilled at the plant spacing of 6″ or 4″ according to the crop spacing needed. Be sure that these holes will not cut the sides of the pipe. These plant holes are to fit the 2″ net pots with the transplants as shown in Figure 12.9.

   Keep all the holes in line at the top edge of the pipe. Do this by snapping a chalk line along the top. Once the holes have been drilled, use sandpaper to eliminate all sharp edges of the holes. Then, glue an end cap at the inlet end and a reduced tee (2″ × 1″) before the end cap at the drain end of each channel.

3. Drill a 1¼″ hole in the reservoir lid at the return end of the NFT channels to allow the 1″ diameter tee and spout to enter the tank for drainage back to it.

4. Make a support pipe to be placed at the top edge of the grow pipes. This is done by using a short piece of 3″ PVC pipe and cutting 2″ holes in it to hold the grow pipes. Be sure that in this case the 2″ or slightly larger holes will cut the side of the support pipe to permit the nesting of the grow pipes. This can be done with a hole saw and then widen the edges with a hacksaw until the grow pipes nest securely. You could also use a short piece of 4″ diameter pipe to do this as the support pipe. The principle here is to support one end

FIGURE 12.9 Net pot in 2″ PVC pipe NFT channel.

of the growing pipe at least 2″ above the lid of the reservoir underneath. This will give adequate slope so that the solution will quickly run back to the nutrient reservoir giving the solution some oxygenation as it falls back into the reservoir. Simply lay the support pipe on top of the reservoir lid at the inlet end of the growing pipe as shown in the plan view of Figure 12.10.

The flow of the nutrient solution should be between 1 and 2 L (¼–½ gallon) per minute. That rate of flow will provide good oxygenation. If you use ¼″ tubing from the pump to the channel insert a small piece of drip line into the end where it is entering the channel to reduce the flow, if necessary.

The final assembly will appear as in the diagram (Figure 12.10).

EBB AND FLOW (FLOOD AND DRAIN) SYSTEMS

This system consists of a growing tray sitting above a nutrient solution reservoir. The principle here is to flood the growing tray with the nutrient solution and then allow it to drain back to the reservoir below. Substrate should be porous such as gravel, pea gravel, expanded clay, or coarse sand. Do not use any fine medium like peatlite, coco coir, or sawdust as these will hold too much moisture and there will be a lack of oxygen to the plants.

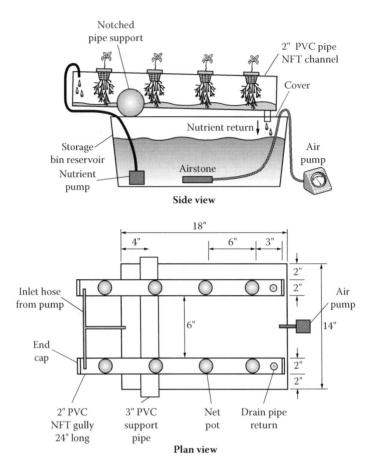

FIGURE 12.10   Side and plan views of simple NFT system. (Drawing courtesy of George Barile, Accurate Art, Inc., Holbrook, New York.)

A submersible pump in the solution reservoir pumps the solution into the tray. The pump is operated by a timer to automatically irrigate several times a day depending on the stage of plant growth and moisture retention of the medium. A sealed fill/drain fitting must be attached to the bottom of the grow tray. When the timer shuts off the pump, the solution will run back to the reservoir below through the pump.

This type of hydroponic system can grow most plants including vine crops of tomatoes, peppers, eggplants, and cucumbers.

SUPPLIES

1. Two storage bins, one for the nutrient tank and the other for the growing tray. The growing tray may be larger than the solution reservoir, but should be shallower. A large grow tray would sit on top of the solution reservoir.

Use at least a 15–20 gallon bin for the reservoir with a depth of 1 ft or more. The grow tray should be wider and longer in order to sit above the solution tank supported by ¾″ square metal tubing spanning the reservoir (Figure 12.11). The depth of the growing tray should be 8–10″. A second option is to have the length and width of the growing tray slightly smaller than that of the reservoir enabling it to nest within the reservoir as shown in Figures 12.12 and 12.13.

2. A frame to support the growing tray may be constructed of 1″ schedule 40 PVC or ¾″ square steel or aluminum tubing that could be cut and bolted together with brackets or if the reservoir bin is strong enough, simply place a few square steel tubing bars across it as shown in Figure 12.11.
3. A submersible pump with ½″ polyethylene tubing.
4. One bulk-head fitting to seal the entrance of the tube from the pump into the bottom of the grow tray. Bulk-head fittings have rubber washers on each side and screw tightly against the tray to seal it from leaking. Alternatively,

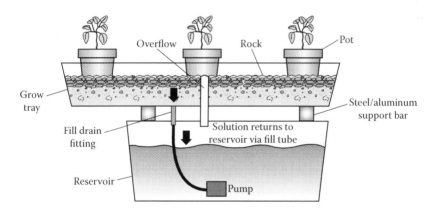

FIGURE 12.11   An ebb and flow indoor unit showing drain cycle with pump off. (Drawing courtesy of George Barile, Accurate Art, Inc., Holbrook, New York.)

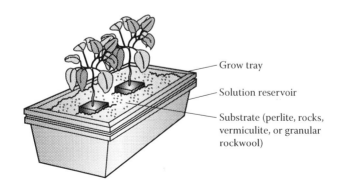

FIGURE 12.12   Alternative ebb and flow system made from two plastic storage bins. (Drawing courtesy of George Barile, Accurate Art, Inc., Holbrook, New York.)

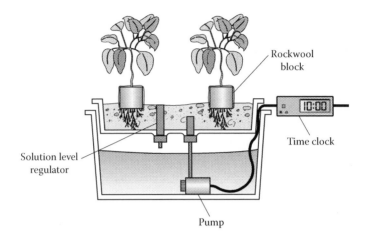

FIGURE 12.13 Piping details of ebb and flow system. (Drawing courtesy of George Barile, Accurate Art, Inc., Holbrook, New York.)

you could use some air-hose fittings available in an aquarium shop. Seal those with silicone rubber.

5. An overflow pipe that regulates the maximum height of the solution entering the grow tray. This pipe should be long enough to regulate the solution level within 1″ from the surface of the substrate in the grow tray and extend below into the solution reservoir to avoid any spillage. It must also be sealed with some type of fitting similar to the pump inlet tube.

ASSEMBLY AND SOLUTION MAKEUP

Assemble the ebb and flow system as shown in the diagrams (Figures 12.11 through 12.13). Probably the most difficult parts to install are the inlet pipe and overflow pipe in the bottom of the grow tray so that a water-proof seal is attained. The construction of the support frame for the grow tray must maintain the grow tray level in all directions above the nutrient reservoir. Be sure to wrap the reservoir with aluminum foil to prevent light from entering. Also, purchase a lid for the storage-bin reservoir. Make an access panel of 2″ × 2″ at one corner of the lid to enable the addition of water to the tank.

When mixing the nutrient solution remove the lid of the tank, pump, and fittings and slide the tank from underneath the growing tray. It is important to construct the frame of the growing tray high and wide enough so as not to restrict movement of the reservoir below for making up the solution. When the plants use up about half of the nutrient solution in the reservoir add water only the first time, and the second time it goes down change the solution. Clean the reservoir well with a 10% bleach solution to disinfect it and then rinse it with clean water before making up the solution.

ALTERNATIVE SYSTEM

This ebb and flow system can be modified to use pots with substrate instead of a full tray of aggregate. To construct this system use a grow tray of 4–6″ deep instead of 8–10″. Place 1″ of aggregate in the bottom of the grow tray to prevent algae growth in the tray. Set three-gallon nursery pots on top of the gravel in the tray as shown in Figure 12.11. Fill the nursery pots with coarse sand or perlite. The rest of the growing tray is made similar to that described earlier with the exception that the overflow pipe is set 2″ high above the base of the tray. The remaining components are the same.

DRIP IRRIGATION SYSTEMS

Drip irrigation is the most widely used and versatile method of hydroponics. Operation is simple with a timer activating a pump to initiate an irrigation cycle. During a cycle of irrigation, the solution is pumped from the nutrient reservoir to the base of the plants through a drip line (Figure 12.14). Drip irrigation systems may recycle the nutrient solution to the nutrient tank or it may leach to waste. In a recirculation (closed) system the pH and strength (EC) of the solution must be checked periodically and adjusted. As a result, the recycled system is more complicated to manage compared to an open system in which the excess solution drainage (leachate) is run to waste. In the nonrecovery design, the nutrient solution is made up and applied to the plants during irrigation cycles with the same pH and concentration without needing adjustments. The nutrient solution may be stored in a large cistern or tank. It can also be mixed as concentrated "stock" solutions that are diluted by

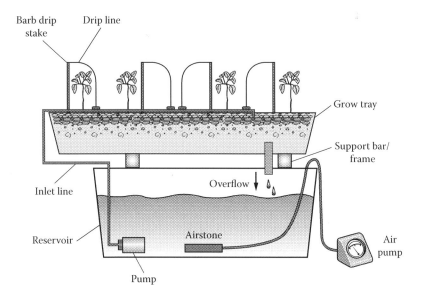

FIGURE 12.14   A drip irrigation system with grow tray of substrate above a nutrient tank. (Drawing courtesy of George Barile, Accurate Art, Inc., Holbrook, New York.)

an injector system upon demand for an irrigation cycle. This is the most common method for large hydroponic operations. For indoor hobby units simply make up a batch of nutrient solution in a storage tank. The larger the tank the less frequent will be its makeup.

Suitable substrates include peatlite, coco coir, perlite, vermiculite, sawdust, rockwool, rice hulls, and expanded clay. Any of these will grow most plants including the vine crops. The basic setup is very similar to that of ebb and flow systems with the exception that the nutrient solution is pumped to the base of the plants at the top of the tray and distributed with a drip manifold to individual drip lines (Figure 12.14). The solution is returned to the tank underneath by a drain pipe sealed to the bottom of the grow tray. The growing tray must be level in all directions to prevent any water accumulation. An option to fill the grow tray with medium is to use pots in the tray. Set them on top of a black weed mat or on about 1″ of gravel or expanded clay to facilitate drainage and to prevent algae growth. The size and number of pots is dependent upon the crop grown. For lettuce and herbs use eight 4″ pots. For vine crops a maximum of two 10″ pots is feasible providing the plants are V-cordon trained in both directions. In this manner, separate the top of the crop sufficiently to get 3.5 sq ft of surface per plant. Place two drip lines in each pot. It is advantageous to include an air pump connected to an airstone placed inside the solution tank to improve oxygenation.

SUPPLIES

1. Two plastic storage bins as described earlier for the ebb and flow system. The grow tray should be 8–10″ deep so that there is adequate space for growing vine crops.
2. PVC or steel pipe framework to support the grow tray above the solution reservoir.
3. Submersible pump with tubing, drip manifold, drip lines, and stakes to hold the drip lines in place.
4. An aquarium air pump placed outside, connected with poly tubing to an airstone (4–6″) in the solution tank.
5. A ¾–1″ diameter drain pipe attached flush with the bottom of the grow tray on one end using a bulk-head fitting or other that gives a complete seal against leaking.
6. A time clock to control irrigation cycles by the pump.

ASSEMBLY

The assembly as shown in Figure 12.14 is very similar to that of the ebb and flow system with the exception of the different placement of the drain pipe and the use of the drip lines to the top of the grow tray. Drip irrigation supplies may be purchased from an irrigation or hydroponic store or online.

Seedlings are started in rockwool cubes, transplanted to blocks (vine crops), and later transplanted again to the grow tray once they are 6″ or so tall. The procedures for growing seedlings are discussed in Chapter 14.

Alternative System

One of the very earliest indoor units using drip irrigation was that of the "tube-in-tube" design as was shown in Figure 1.2 of Chapter 1. A fish aquarium air pump is mounted on one edge of the growing tray with a ¼″ diameter poly hose connecting it to a slightly larger diameter hose in the nutrient solution of the reservoir underneath. The key to success here is to use a slightly larger hose to join about 1″ insertion of the smaller diameter hose allowing the air to suck in the nutrient solution as it enters the larger hose. Connect the larger tube to the smaller one with a pin. The larger hose must be loose enough for the solution to enter at the union as the air is bubbled up through the larger hose. The movement of air drags the nutrient solution into the larger tube and raises it up into the grow tray above. The larger hose enters the grow tray and lies on the top of the substrate. This area of the hose above the substrate has small holes drilled into it every 2″. The holes must be no larger than ¹⁄₁₆″ in diameter so that all of the nutrient will be carried along its length on the top of the medium.

To support the grow tray without a tray support frame do as follows: Purchase the grow tray a few inches smaller in length and width than the reservoir. Drill ¼″ diameter drainage holes in the bottom of the grow tray 3″ × 3″ spacing to within 2″ of the outside perimeter. Do not use a lid on the reservoir below. Place two or three ½″ square aluminum- or chrome-plated supports across the top of the nutrient tank and set the grow tray on top of those bars in a position between the drainage holes. To use a lid on the lower tray, it is possible if it has a lip that seals the bin around the edges. However, to get good drainage position the upper grow tray on top of the lid of the lower reservoir and drill holes through both the bottom of the grow tray and the lid of the reservoir at the same time in the identical same positions. Glue or silicone a few guides, using small plastic tubing, positioned at the corners of the grow tray and adhered to the lid of the reservoir. In that way, when you place the grow tray on the top of the reservoir, the drainage holes will line up. Refer to Figure 1.2 to follow the assembly of the drip systems.

The air pump can be activated with a time clock for the use of fine substrates in the grow tray or for coarser material such as rocks or pebbles allow the pump to run constantly. The air pump not only moves the solution from the reservoir below, but also adds oxygen to the solution as it moves up to the grow tray.

For this form of drip system, you may use expanded clay, pea gravel, coarse sand, perlite, vermiculite, or rockwool granules. Do not use peatlite mix or coco coir due to excessive moisture retention. You could use rice hulls as they have lower water retention. Coarse vermiculite and perlite are the best substrates for this system. If you wish to use pea gravel or expanded clay add a ½″-thick layer of peatlite or coco coir layer on top to achieve capillary sideways movement of the solution.

ROCKWOOL SYSTEM

Rockwool is a common substrate used in hydroponic culture and it is available at hydroponic stores. As discussed in Chapter 14, the seedlings are started in

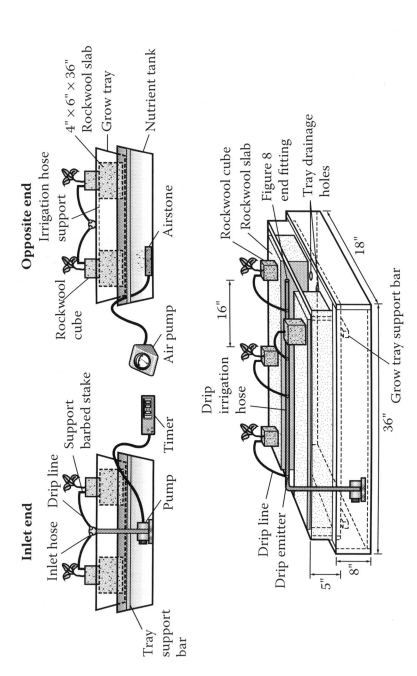

FIGURE 12.15 Rockwool slab system with drip irrigation and a nutrient tank below. (Drawing courtesy of George Barile, Accurate Art, Inc., Holbrook, New York.)

rockwool cubes that are transplanted to rockwool blocks. From there, they are transplanted to the rockwool growing system using rockwool slabs 4″ thick by 6″ wide by 36–39″ long. The simplest method of construction for an individual unit is to purchase a plastic bin of minimum length of 36″ by 18″ wide by at least 8–10″ deep (Figure 12.15). A greater depth is fine as that will increase the overall volume of the tank and reduce the frequency of adding water and/or changing the solution. If you can find this size of container, a second grow tray above will not be necessary providing the tray has a locking lid that positions the lid top below the sides of the bin. If such dimensions are not available use two bins as described for the drip irrigation system.

Use the lid as the support tray for the rockwool slabs. With the preceding dimensions of the bin, the system will fit two slabs. Feeding of the plants to their bases is by a drip irrigation system operated by a timer. As described for drip irrigation systems, position the end of the drip line with a special barbed stake guide. There are several differences with this drip system compared to the ones described earlier. The pump is connected to PVC pipe that is converted to ½″ black poly hose using a barbed adapter at the entrance to the slabs above the lid. The black poly line runs along the length of the lid between the two slabs. Drip lines are attached to the black poly line using 0.5-gallon per hour pressure compensating emitters punched into the line. Each emitter is attached to a 16″ length of drip line of 0.160–0.220″ diameter.

With vine crops such as tomatoes, peppers, cucumbers, or eggplants, each slab will contain three plants so a total of six drip lines are required, one to each plant base. The end of the black poly hose is bent over and secured with a figure "8" adapter or a short 3″ piece of 1″ diameter PVC pipe. The key to success with this number of plants is the V-cordon training in both directions to provide 3.5 sq ft per plant at ceiling height. Plant only one cucumber plant per slab.

SUPPLIES

1. One plastic storage bin with a lid measuring 36″ × 18″ × 8–10″ or more in depth.
2. Submersible pump of minimum capacity of 80 gallons per hour (gph).
3. Compensating emitters of 0.5 gph.
4. Drip line—0.160–0.220″ diameter, 10 ft.
5. Barbed stakes: six.
6. Figure "8" end stop or 3″ of 1″ diameter PVC pipe.
7. Schedule 40, ¾″ diameter PVC pipe: 2 ft.
8. Male adapter to fit pump outlet diameter, a reduced bushing from ¾″ diameter to the diameter of the pump outlet.
9. Fittings: ¾″ PVC 90° elbow, ¾″ × ½″ slip by thread bushing, ½″ barbed adapter (thread-barb), 1″ hose clamp.
10. PVC glue and cleaner
11. An aquarium pump, poly tubing, and 6″ airstone to aerate the solution.
12. A 24-h time clock with 5–15 min intervals, no greater.
13. A poly punch tool to make the holes in the black poly hose for the emitters.

ASSEMBLY

First assemble the drip hose and emitter system. Glue the various fittings, punch the holes for the emitters (six for vine crops) evenly along the length of the black poly hose starting at 3″ from the PVC adapter and no closer than 6″ to the end. Push in the emitters and attach the drip lines with the barbed stakes at their ends. Then, make up the PVC piping from the pump to the black poly lateral hose. Use Teflon tape on all threaded fittings. Refer to the diagram for details.

With the air pump attach a section of poly tubing long enough to reach the inside of the reservoir and there connect to the airstone.

Make ¼″ drainage holes on the top of the lid for recycling the solution leachate from the slabs to the reservoir below. Make two rows of these holes about 4″ apart going down the center of the lid and within 1½″ of the edge. Drill the two rows of holes 4″ apart and stagger their positions (Figure 12.16). At one end make holes 4″ apart to intercept any runoff that does not go down the center of the lid.

Drill a hole large enough to fit the PVC pipe from the pump in the middle of the opposite end to the drain holes, 3″ from the edge of the reservoir. Construct a little stand about 1″ high at one of the corners of the lid to support the air pump. Drill a ½″-diameter hole for the air hose to enter the reservoir to connect to the airstone. Cut a 3″ × 3″ square access panel in the lid at the middle or corner of one end to permit the addition of water to the reservoir. Refer to the diagrams for details (Figures 12.15 and 12.16).

If you cannot find a storage bin 36″ long, design the system as explained earlier, but cut the rockwool slabs shorter to fit the lid dimensions. It may be possible to grow only four vine crops in such a smaller unit. If you cut the end of a slab staple it closed.

When growing vine crops that are trained vertically in a small hydroponic unit support them above with a wire and space this wire to give adequate light to the entire plants as explained in Chapter 24. V-cordon train them away from the grow tray. The support cable must be at least 30″ apart at the top of the mature plants. In your basement or a spare room use ceiling hooks to support the plant strings.

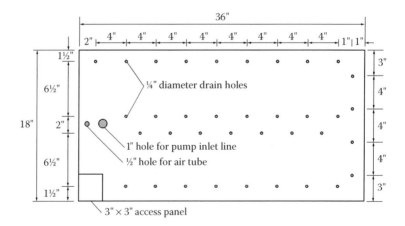

FIGURE 12.16 Drainage hole positions in lid of nutrient tank. (Drawing courtesy of George Barile, Accurate Art, Inc., Holbrook, New York.)

COCO COIR SYSTEM

This system is the very same as the rockwool one, but instead of using rockwool slabs, use coco coir slabs. The dimensions of the coco coir slabs are similar to those of rockwool. The biggest difference in cultural techniques is that coco coir needs less frequent irrigation cycles compared to rockwool. Generally, two to three cycles per day are sufficient. You need a 15% leachate during each irrigation, whereas with rockwool the leachate should be 20%–25%.

AEROPONIC SYSTEM

An aeroponic system is one of the most highly technical methods of hydroponics and most risky should a loss of power occur to delay irrigation cycles. In this system, the plants are supported in the top (lid) of the reservoir with their roots suspended below in the air space between the solution and top cover. The roots are misted every few minutes with nutrient solution. The submersible nutrient pump is controlled by a timer that runs the pump for a 1 min and 4 min off on a 24-h basis.

The nutrient reservoir should be of 15–20 gallons with a depth of at least 12″. It must come with a lid that secures the sides of the bin. This system is best for small plants such as lettuce, basil, arugula, and herbs, but with care and experience it could grow vine crops. All of the crops are started in rockwool cubes and transplanted to 2″ mesh pots that are placed in 1¾″ diameter holes on the top of the bin lid. For lettuce, basil, and arugula use 6″ × 6″ spacing of holes, and for other herbs you may use 4″ × 4″ spacing. Stagger the position of the holes if possible. The exact layout of the plant holes depends upon the dimensions of the storage bin. For example, a bin 16″ wide by 20″ long by 12″ deep, as shown in Figure 12.17, has a volume of 17 U.S. gallons. It would hold eight plants (four per row) in two rows spaced 8″ apart. The spacing within each row is 2½″—5″—5″—5″—2½″.

SUPPLIES

1. A 15–20-gallon storage bin (2–2½ cu ft) with lid.
2. A high pressure submersible pump.
3. PVC pipe of ¾″ diameter for the mist nozzle distribution pipe and the support pipe frame.
4. Various PVC fittings for the mist distribution pipe.
5. An accurate 24-h timer with minute intervals. Alternatively, use two timers, one of 24-h intervals and the other of 1 h with 1-min intervals. Place these in series.

ASSEMBLY

The pump and all piping are set within the nutrient reservoir. The mist jets mounted in the PVC header pipe is located above the level of the nutrient solution. Assemble the piping and mist jets first then place them in the nutrient reservoir. A supporting

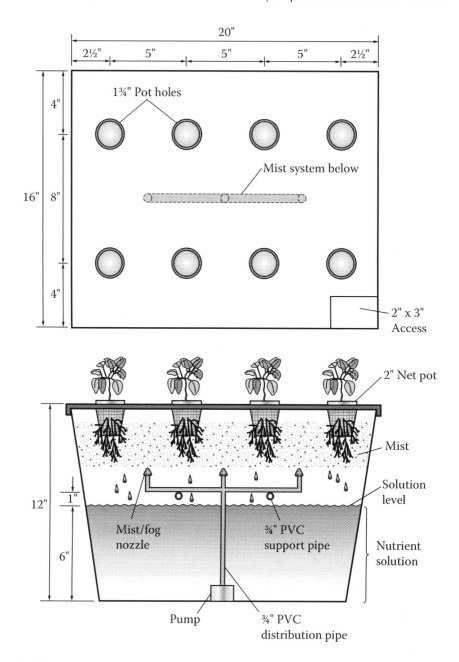

FIGURE 12.17 Aeroponic simple system showing mist jets. (Drawing courtesy of George Barile, Accurate Art, Inc., Holbrook, New York.)

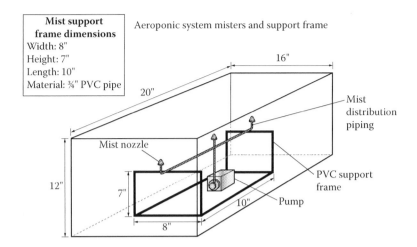

FIGURE 12.18 PVC frame supporting mist jets in nutrient tank. (Drawing courtesy of George Barile, Accurate Art, Inc., Holbrook, New York.)

frame of PVC pipe is constructed to support the mist-jet manifold above the level of the nutrient solution.

Cut 1¾″ diameter holes in the reservoir lid at spacing according to the nature of the crop as described earlier. Make a 2″ × 3″ access panel at one corner of the lid to add water or nutrients. The mist nozzles should be located about 4″ below the top of the lid. The maximum level of the solution is to be 6″ to permit adequate air space above for the mist distribution and the aeration of the plant roots. These details are outlined in Figures 12.17 and 12.18.

ALTERNATIVE DESIGN

Use two bins, a lower nutrient reservoir and an upper growing tray. The growing tray should be several inches smaller in length and width from the reservoir so that it will sit flat on the cover of the reservoir. The growing tray should be about 6″ deep with a cover. It sits on top of the lid of the nutrient tank. Install a ½″ diameter drain pipe at the center of one end of the grow tray to return the solution to the tank below. Seal the drain pipe to the bottom of the grow tray as explained earlier for other units.

Use ½″ high-pressure tubing from the submersible pump to the base of the grow tray. As it enters the grow tray use a rubber washer to seal it. Connect the tubing to a ½″ barbed tee as it enters the growing tray. With a short piece of tubing connect both sides of the tee to ½″ PVC using barbed-PVC adapters as shown in Figure 12.19. This will allow the growing tray to be detached during cleaning between crops. In the grow tray, the PVC pipe is branched with the tee to form two mist nozzle manifolds. Insert the mist nozzles at about 6″ spacing along the manifold depending upon the grow tray dimensions. The misting assembly sits on the base of the grow tray as shown in the diagrams of Figure 12.19. Cut holes in the grow-tray cover for the plant

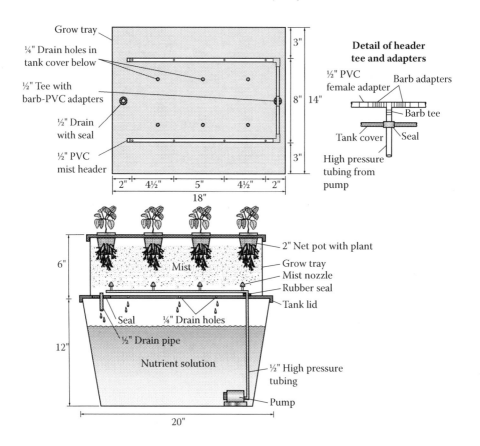

FIGURE 12.19 Alternative design using two bins with details of the aeroponic system. (Drawing courtesy of George Barile, Accurate Art, Inc., Holbrook, New York.)

pots as described earlier. Use Schedule 40 or 80 PVC pipe that has thick walls so that the mist nozzles can be attached to the pipe.

Drill ¼″ holes in the lid of the nutrient tank about 5–6″ apart for drainage back to the solution tank of any solution that may leak from the grow tray.

These types of aeroponic units are available at hydroponic shops and online (refer to Appendix).

13 Large Indoor Hydroponic Units
Designs and Construction

Chapter 12 exemplified individual unit systems. This chapter shows how to construct a series of pots, gutters, slabs, and so on that drain to a central reservoir. These growing systems are more complex in their construction due to the need for supporting structures for the growing trays and more plumbing to connect them to the feeding and draining systems. Some of the systems are more suitable to low-profile plants such as lettuce, arugula, basil, and herbs, while others are more versatile and can also grow vine crops such as tomatoes, peppers, cucumbers, and eggplants.

MULTI-POT OR GROW TRAY WICK SYSTEM

This system is the same principle as the single-pot wick design, but instead of just one pot a series of pots are supported in the lid of the nutrient solution reservoir (Figure 13.1). This system is for low-profile crops. Use 4″ diameter plastic pots to contain perlite and/or vermiculite as the substrate. These pots are supported by the lid of the nutrient tank. Drill approximately 3¾″ diameter holes in the lid at 6″ × 6″ spacing. Be sure to measure the diameter of the pot directly below the lip so that the top lip section of the pot remains above the lid.

A second method is to place a grow tray on top of the nutrient solution tank and extend wicks from the solution tank through the base of the grow tray into the substrate of the grow tray where the plants are positioned, as was presented in Figure 12.3.

MATERIALS

1. Storage bin with a snap-on lid
2. Cotton or fibrous nylon rope
3. Four-inch diameter plastic pots
4. Perlite and vermiculite mixture at half and half of each
5. Air pump and 4″ airstone

ASSEMBLY

Drill the holes for the pots in the lid staggering their positions. If a large storage bin greater than 12″ × 18″ is used, reinforce the lid by placing a ¼″-thick plastic sheet on top and drill the holes at the same time through it and the lid to line them up

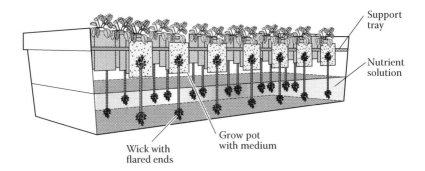

FIGURE 13.1 Multi-pot wick system using a large plastic storage bin. (Drawing courtesy of George Barile, Accurate Art, Inc., Holbrook, New York.)

correctly. Without this additional support, the thin lid of the reservoir storage bin will sag and possibly break with the weight of the medium and plants.

Make a 2″ square access panel at one corner of the lid to add water and/or nutrients. This can also be the entrance of the poly tube from the air pump to the airstone within the reservoir.

Flare the ends of the rope to form the wick and bury it into the medium of the pot about one-third. If using a grow tray of medium, locate the wicks below the plant sites and drill through the grow tray at those positions. Make the wicks long enough to reach within ½″ of the bottom of the solution reservoir. Add raw water to the nutrient reservoir, place the lid on top, and add the pots. Seed the pots in place in the system and water the pots from above to initially moisten the substrate. Use a half-strength nutrient formulation after the plants have their first true leaves unfolding. About 10–14 days later, make up the full-strength formulation solution.

WATER CULTURE (FLOATING) SYSTEM

This culture is most suited to low-profile leafy crops such as lettuce, spinach, basil, arugula, and various herbs. The design is similar to the small-scale unit, with the exception that beds (sometimes called raceways) are larger. Cut the boards from 4 ft × 8 ft × 1″ thick "Roofmate" Styrofoam. The raceway can be constructed of 2″ × 10″ treated and painted cedar planks and lined with a 20-mil vinyl swimming pool liner (Figure 13.2). The raceway may be constructed on the floor or as a raised bed. The raised bed form is better as it can drain by gravity to an underlying cistern (Figure 13.3).

MATERIALS

1. Lumber of 2″ × 10″ dimension. Length to suit your specific needs, but the raceway must be a multiple of 2 ft in length to fit the boards.
2. Swimming pool vinyl liner of 20-mil thickness.
3. Vinyl cement.
4. Various PVC pipes (¾″ schedule 40) and fittings.

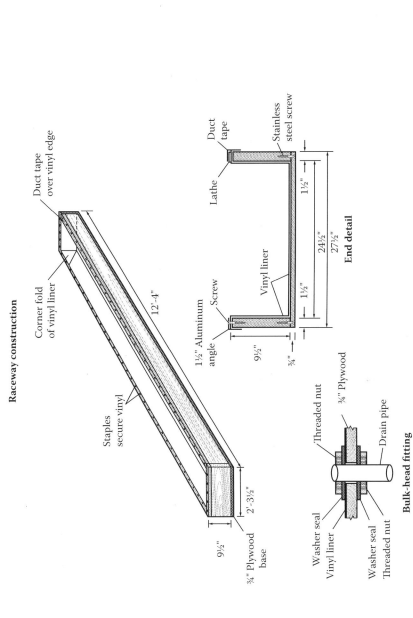

Raceway construction

Duct tape over vinyl edge

Corner fold of vinyl liner

Staples secure vinyl

12'-4"

2'-3½"

9½"

¾" Plywood base

End detail

Duct tape

Lathe

Stainless steel screw

Vinyl liner

Screw

1½" Aluminum angle

1½"

24½"

27½"

1½"

9½"

¾"

Bulk-head fitting

Threaded nut

¾" Plywood

Drain pipe

Washer seal

Vinyl liner

Washer seal

Threaded nut

FIGURE 13.2 Construction of wooden raceway for raft culture. (Drawing courtesy of George Barile, Accurate Art, Inc., Holbrook, New York.)

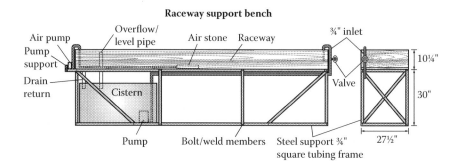

FIGURE 13.3 Support frame with cistern underneath and piping details. (Drawing courtesy of George Barile, Accurate Art, Inc., Holbrook, New York.)

5. Lathes 1″ × ¼″ thick or 1½″ × 1½″ aluminum angle.
6. An air pump and air stones are available from fish aquarium stores or aquaculture suppliers.
7. Submersible pump with a timer.
8. Nursery landscape weed mat if this is outside or in a greenhouse.
9. 6–8″ nursery staples to secure the weed mat.
10. 50–100-gal cistern tank. The size depends upon the length of the raceway. For a short raceway up to 20 ft, a 50-gal cistern is adequate.
11. Galvanized or chrome plated ¾″ square steel tubing for a support frame.
12. White oil-based paint.

Assembly

If the system is located inside the house, a weed mat is unnecessary. The raceway sits above a solution tank so it must be raised up on a supporting frame (Figure 13.3). Construct the raceway before the support frame.

Make the inside dimensions of the raceway frame ½″ wider and 1″ longer than the total dimensions of the boards. For example, to construct a raceway 12 ft long by 2 ft wide, make the inside dimensions 12 ft 1″ (145″) long by 2 ft ½″ (24½″) wide to allow for some free play and the thickness of the vinyl liner. If the raceway is to be located on a frame, make a bottom of ¾″ plywood. Attach the frame and bottom with stainless steel screws to avoid any corrosion and glue all joints for added strength as water is very heavy. Cut a 1½″ diameter hole at the bottom of one end of the raceway, about 4″ from the end in the center. Paint the inside and outside of the raceway framework with an oil-based paint.

Line the raceway with 20-mil vinyl liner. Place it tightly against the bottom and all corners to get a smooth flat surface. At the corners fold it similar to wrapping a parcel and glue the wrapped corners. Bring the vinyl over the edge of the wooden frame and secure it under a lathe on top around the perimeter with aluminum nails or stainless steel screws. Alternatively, staple the vinyl to the top edge of the frame and cover the staples with ducting tape bringing the tape over the edge to make a very nice finish as shown in Figure 13.2. To make a very nice finish, screw an aluminum angle on top of the vinyl around the top edge of the raceway. You will have to make a small slit at the top of each corner to permit the vinyl to lay flat at that point. Cut the excess vinyl flush to the wooden frame. Where the drain hole was cut, hold the vinyl tightly and make a cross cut in the vinyl to within ¼″ of the width of the hole. Install a bulk-head fitting through the vinyl and hole gluing the vinyl between the bulk-head fitting seals. Caulk with silicone rubber. The bulk-head fitting must give a complete seal to avoid any leakage of water between the liner and plywood bottom of the frame.

Construct a steel tubing frame to support the raceway (Figure 13.3). The frame must keep the raceway about 30″ above the floor to permit the cistern to fit underneath. Locate the cistern under the drain pipe of the raceway. The support frame must be level in all directions to get good drainage from the raceway. Locate an air pump at one corner of the raceway frame and run a poly hose down the center connecting to a 6″ air stone.

Place a 1½″ diameter by 9″ long PVC pipe in the drain to maintain 8″ of solution in the raceway. Do not glue this pipe. This is the overflow pipe to re-circulate the nutrient solution to the cistern.

Cut a notch in the top of this pipe so that flow from the raceway will not be impeded by the boards floating on top of the solution.

Install a submersible pump with a ¾″ outlet and plumb it with fittings and ¾″ PVC schedule 40 pipe from the pump underneath and attached to the metal support frame going to the opposite end from the drain. Then, with elbows, extend the pipe over the end of the raceway to enter the top of the raceway as an inlet for the solution (Figure 13.3). Put a ball valve at the vertical side to regulate the flow. This keeps a constant flow of solution aerating and replenishing the solution in the raceway for the plants.

The next step is to make the boards. Cut the boards 6″ × 2 ft, 1 × 2 ft, 2 × 4 ft, or 4 × 4 ft long exactly depending upon the raceway width. Use a very sharp kitchen knife or gyproc knife guiding it with a straight edge. Alternatively, a saw with fine teeth will do the job. Smooth the edges with fine sand paper. Drill ¾″ diameter holes at 6″ centers for lettuce, arugula, and basil with 4″ centers for small herbs. For 6″ centers start the holes 3″ from the edge. For 1 × 4 ft boards make two rows of four holes forming eight plant sites per board. The hole pattern across the raceway is 3″–6″–6″–6″–3″ (four holes) and the other way 3″–6″–3″ (two rows). For 4″ centers, start 2″ from the edge and space 4″ apart to get a total of six holes within the row and in the other direction start again 2″ in from the edge to form three rows. They are not staggered in their position. After making the holes, smooth the edges with sand paper. Refer to the diagrams of Figure 13.4 for the board dimensions and hole patterns.

ALTERNATIVE SIMPLIFIED SYSTEM

In this design we can eliminate the nutrient reservoir below the raceway. You can even eliminate the supporting frame if you wish to construct the raceway on the floor. Simply set the raceway on top of 1–2″-thick Styrofoam for insulation from a cold basement floor.

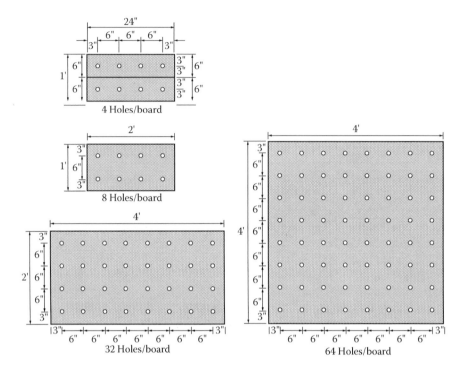

FIGURE 13.4 Details of hole spacing in raft culture boards. (Drawing courtesy of George Barile, Accurate Art, Inc., Holbrook, New York.)

Construct the raceway itself of dimensions either 2 or 4 ft wide by whatever length is convenient for your space. Since plywood sheets come as 4 ft wide, make the overall outside bed dimension a maximum of 4 ft. With the framing of 2″ lumber, which is dressed dimension to 1⅝″ thick, the sides will use 2 × 1⅝″ = 3¼″. The Styrofoam sheets will have to be cut to a width of 48″–(3¼″ + ½″ of play) = 44¼″ or about 44″ to allow sufficient free play between the sides and liner. The boards may be 2 ft or 4 ft × 44″.

Place the raceway at the floor level on Styrofoam insulation or make a frame as described earlier. However, no drain hole for returning the solution to a reservoir tank is needed. Make up the nutrient solution directly in the raceway and keep it mixed and aerated using an air pump and tubing to air stones as explained earlier. This design eliminates the complexity of returning the solution to a nutrient tank, thus avoiding the drainage and irrigation systems requiring a pump, and so on.

Check the pH and electrical conductivity of the solution at least once a day and add a small portion of solution (up to about 10% of the original makeup) every few weeks, if necessary. Change the nutrient solution once every few months by pumping it out with a sump pump and cleaning the raceway with a 10% bleach solution.

The plants that were growing in the raceway can be taken out in situ with the boards and kept moist by stacking the boards together root-to-root. You may also mist the roots with water to keep them from drying while cleaning the raceway. As soon as you finish cleaning the raceway and start filling it with water, place the boards with their plants still in place back into the raceway and then make up the nutrient solution.

NFT SYSTEM

This is the most common method of hydroponics for low-profile crops such as, lettuce, basil, arugula, bok choy, and spinach. The growing channels may be constructed from 2″ PVC pipes as shown in Figures 12.9 and 12.10 or it is best to purchase special NFT channels from a hydroponic store. The commercial NFT channels have either ridges on the inside bottom of the channels or a sloped bottom with or without ridges to direct water from flowing around plant roots when the seedlings are transplanted (Figure 13.5). There are two types of channels: a one-piece fixed top and a two-piece with a removable top to facilitate cleaning between crops (Figures 13.5 and 13.6). The covers have plant site holes of correct diameter and spacing for the crop you wish to grow.

The NFT system, regardless of whether it is PVC pipe or other channels, must be located on a frame above a solution reservoir. The bench must have a 3% slope back toward the reservoir. The channels must not exceed 12 ft in length to prevent a temperature rise and oxygen loss in the nutrient solution as it travels along the channel. The benching should be at waist height, about 30″.

Many small NFT systems of various sizes and number of channels are available from hydroponic stores and online (refer to the Appendix). The following materials list is for a system of four channels.

MATERIALS

1. NFT channels 12 ft in length or 2″ PVC pipe 12 ft (8 ft length shown in Figures 13.7 through 13.9).
2. PVC fittings—elbows, tees, reduced bushings, ball valves, and so on.
3. PVC cleaner and glue.
4. PVC piping—1″ and ¾″ diameter schedule 40.
5. A submersible pump with 1″ or ¾″ outlet. Flow rate must be 2 L per minute per channel.
   For four channels, a pump of at least 3–4 gal per min (gpm) is needed with a head (lift) of 6 ft.
6. Drip irrigation tubing—about 10 ft.
7. Square steel tubing ¾″ by ¾″ or lumber 2″ × 2″ treated and/or painted white.

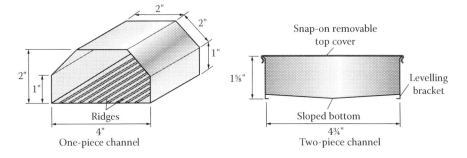

FIGURE 13.5 Commercial NFT gutters. (Drawing courtesy of George Barile, Accurate Art, Inc., Holbrook, New York.)

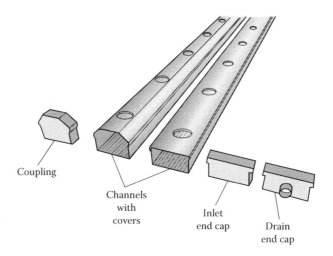

Coupling

Channels
with
covers

Inlet
end cap

Drain
end cap

FIGURE 13.6 Various NFT components and channel configurations. (Drawing courtesy of George Barile, Accurate Art, Inc., Holbrook, New York.)

8. A plastic 50-gal nutrient reservoir, preferably opaque in color with a cover. The nutrient tank may be smaller for shorter and less channels per system. For example, two 10- to 12-ft channels can be served with a 20-gal reservoir.

ASSEMBLY

A small system is discussed first followed by further details of larger, more complex systems. Starting with two 8-ft channels and a 20-gal reservoir, the channels can drain at the lower end directly into the nutrient reservoir at the lower end. The first step is to construct the supports for the growing channels. With two channels that are spaced 6–8″ apart, set the drain ends on the edge of the nutrient reservoir (Figure 13.7).

The support frame can be constructed the same for either 2″ PVC channels or purchased NFT channels. Construct one A-frame "sawhorse" structure 3″ higher than the height of the nutrient tank, for the upper inlet end of the channels. It should be about 2 ft long. If you prefer not to rest the lower end on the nutrient tank, make a second A-frame that is several inches higher than the nutrient tank and make the inlet support frame 3″ higher than the lower one. These support frames can be constructed of wooden 2″ × 4″ lumber and painted white. They can also be made of ¾″ square steel tubing or PVC pipe. But, when using PVC make the frame somewhat different from wood or steel tubing. Use 1″ PVC schedule 40 pipe. First, make a rectangular base 18″ × 24″. In the middle of the 18″ sides attach a tee. From there glue a riser of 14–18½″ depending on the height of the nutrient tank as the taller support must be 5″ above the tank height if the lower support at the tank end is 2″ above the tank. This gives a slope of 3″ for the growing channels. Use 1″ tees at the top of the riser and connect a horizontal pipe between the two tees. This completes the supports as shown in Figures 13.7 and 13.8. Another one exactly the same for the middle

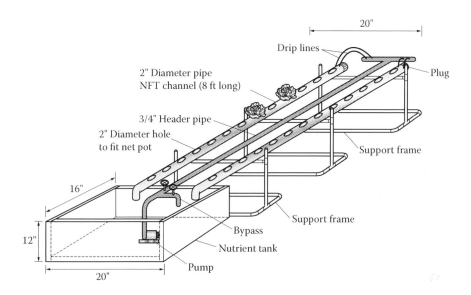

FIGURE 13.7 NFT pipe system. (Drawing courtesy of George Barile, Accurate Art, Inc., Holbrook, New York.)

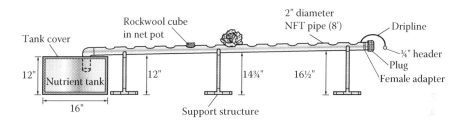

FIGURE 13.8 NFT pipe system side view showing support structures. (Drawing courtesy of George Barile, Accurate Art, Inc., Holbrook, New York.)

is needed to prevent the channels from sagging with the weight of the solution and plants. Make this one 1½" shorter than the highest one at the inlet end.

The piping from a submersible pump in the nutrient tank to the upper inlet end of the channels is as follows. Attach a ¾" pipe to the pump with a threaded male adapter. Make this riser long enough to reach up through the tank lid and to the top of the support frame. Just above the tank cover, install a return bypass line with a ball valve to regulate the flow of solution to the inlets of the channels. In this way, if the pump has more capacity than needed, the flow is shunted back to the tank. From there use a 90° elbow to attach the pipe going on top of the support frames to the inlet end of the channels. At the inlet end attach a tee and make a header going each way.

Glue a cap on one end of the header and a threaded female adapter to which a plug can be placed to permit cleaning of the pipe. This header should be about 18–20" long to span between both growing channels. From it use a special grommet seal and insert several drip lines for each channel. Two drip lines enter the top of each

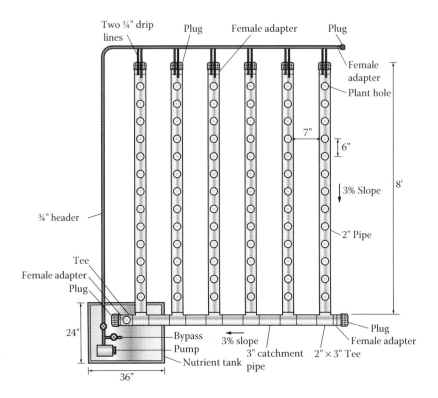

FIGURE 13.9 Details of NFT pipe system with multiple growing pipes. (Drawing courtesy of George Barile, Accurate Art, Inc., Holbrook, New York.)

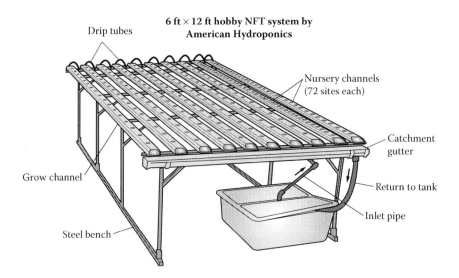

FIGURE 13.10 Details of a 6 ft × 12 ft hobby NFT system. (Drawing courtesy of George Barile, Accurate Art, Inc., Holbrook, New York.)

growing channel by drilling snug holes. Alternatively convert to ½″ black poly tubing with barbed adapters at the tee and punch holes for drip lines at the grow pipe positions. The solution flows constantly so a timer is not needed for the submersible pump. Refer to Figures 13.7 and 13.8 for details of the design.

If you wish to make the grow channels from 2″ PVC pipes, you must drill the plant site holes similar to that mentioned in Chapter 12 for simple units. Make the holes 2″ in diameter as the mesh pots have a ¹⁄₁₆″ lip around their tops as shown in Figure 12.9. This will prevent the mesh pot from falling into the pipe. Locate the holes in a straight line at 6″ centers for lettuce, arugula, bok choy, and basil. To hold the growing pipe in position on the support frame, use plastic electrical straps around the 2″ PVC growing channels securing them to the 1″ PVC support frame. Glue a 2″ female adapter on the ends of the growing channels and fit them with a threaded plug to enable access for cleaning. A multi-pipe NFT system is set similar to the two-pipe system with the exception of plumbing the drain ends of the grow pipes into a 3″ diameter catchment pipe that returns the solution to the nutrient tank (Figure 13.9). The irrigation system is the same as described earlier as appears in Figures 13.7 and 13.8.

LARGER NFT SYSTEM

The next are steps to build a somewhat larger and more complex system. The materials list is basically the same, but with more channels and a larger benching frame. The following is a design similar to that available from American Hydroponics (see Appendix). The system has eight production channels and two seedling channels to grow 144 lettuce, basil, arugula, or bok choy. It occupies an area of 6 ft × 12 ft. Construct the bench framework first using ¾″ square steel as shown in Figure 13.10. The bench is 34–38″ high. This gives a slope of 4″ in 138″ or about 3%. Since the channels are exactly 12 ft long, make the bench slightly shorter at 11½ ft (138″) to allow the channels to extend out from the bench. There are a total of four cross members placed at 46″ centers. Set these on a base extending the 11½ ft on each side. Put braces between each corner at the top of each cross member as shown in the diagram. It would be best to weld the entire structure, but if that is not possible, drill holes and secure each member with bolts.

Each of the eight growing channels has 1¾″ diameter holes spaced at 8″ centers to permit 18 plant sites per channel as shown in Figure 13.11. The two nursery channels have the same size of holes spaced at 2″ centers to fit 72 seedlings per channel. The production gullies house 144 plants (8 × 18 = 144). The gullies come with a cover and are ready to set up with the irrigation system. They drain into a closed collection pipe that conducts the solution back to the nutrient tank.

There are many sizes and types of NFT systems available from hydroponic stores and online. If you do not want the work of constructing one, I recommend purchasing a complete unit (see Appendix). You may also buy NFT gutters of various lengths (4 ft, 6 ft, or 8 ft).

To grow vine crops use wider channels or larger PVC pipes of 4–6″ diameter. Wider NFT channels of 4″, 6″, and 9″ with plant site holes at 8″ for the 4″ and 6″ wide channels and at 12″ for the 9″ channel are available. Plant hole diameters are

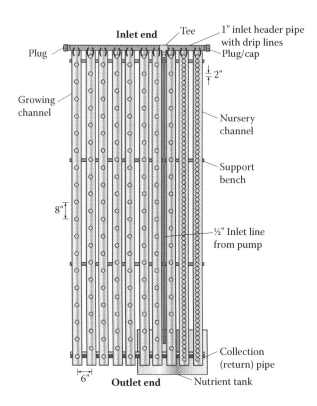

FIGURE 13.11   Plan of gullies on a bench with hole locations for nursery and growing channels. (Drawing courtesy of George Barile, Accurate Art, Inc., Holbrook, New York.)

offered for 2″, 4″, and 6″ net pots. The larger net pots can be filled with substrates as expanded clay particles or granular rockwool.

For vine crops use 9″ wide channels with 12″ spacing for 6″ net pots with a substrate. Whatever substrate or sizes of pots are utilized, begin the seedlings in the conventional method of rockwool cubes and transplant to the net pots. When using a substrate in large net pots, this is really a combination of NFT and another culture.

The following is a smaller NFT system using four channels of 8 ft in length. Purchase the channels with end caps and covers from a hydroponic supplier or make your own from 2″ PVC pipes as described earlier. A support frame and inlet and return piping will have to be constructed. The system described is for lettuce, arugula, basil, bok choy, and herbs in 2″ pipes and vine crops in 4″ diameter pipes.

MATERIALS

1. Two 8 ft 2″ PVC pipes and two 8 ft 4″ PVC pipes.
2. Fittings: two 2″ caps, two 4″ caps.
3. Construct the support frame of 1½″ PVC pipe or of ¾″ square steel tubing.
4. Return pipe of 3″ PVC pipe schedule 40.

5. Five 3″ tees, two 3″ female adapters, two 3″ thread plugs, two 3″ × 2″ reduced bushings, two 4″ × 3″ reduced bushings.
6. Twenty feet of ¾″ diameter PVC for inlet main and header with fittings for attachment to the pump including bypass and ball valve.
7. Submersible pump of ¾″ outlet of at least 4 gpm capacity.
8. Timer 24-h cycle with 1 min intervals.
9. A 20-gal plastic storage bin with a lid for the solution reservoir.
10. Twenty feet of drip line, with grommets to attach to the inlet header.

Assembly

Start with the support frame. Construct it of 1½″ PVC pipe or ¾″ steel tubing. Make the frame dimensions 90″ by 54″ wide by 24″ high on the inlet end side and 21″ at the drain end as shown in the diagram (Figure 13.12). The channels sit on the 48″ wide side. Using a male adapter and several 90° elbows connect the ¾″ inlet line from the pump to a header. Secure the header with strapping underneath the support frame. Install a bypass pipe with a valve above the lid of the reservoir to permit regulation of flow to the inlet pipe. Install the drip lines in the header and into the channels (two per channel). Alternatively, convert to ½″ poly hose for the header into which the drip lines are inserted. They can be inserted without emitters to achieve a minimum flow rate of 1–2 L per minute (¼–½ gal/min).

At the drain ends of the channels assemble a 3″ PVC collection pipe that returns the solution to the reservoir. This is held up by the bench frame with brackets. Two methods for the collection pipe and drainage of the channels are feasible. One is to leave the ends of the growing pipes open and allow them to empty into the collection pipe at open slits 2″ wide. If you use NFT commercial channels attach a 1″ diameter drain pipe to the bottom of the gullies at their drain end and drill 1½″ holes in the collection pipe at the positions of the drain pipes from the gullies. The difficulty doing it this way is to seal the joints with glue or silicone rubber to the NFT gully. However, the advantage is that light can be excluded from entering the collection pipe, which will prevent algae growth. If using PVC pipe channels, glue them into the 3″ PVC header using tees. With this method end the collection pipe with threaded plugs to enable cleaning. Then, place one tee in this collection manifold above the tank to conduct the solution to the tank as shown in Figure 13.12. If the former method of an open gully end into the collection pipe is used, cover the outlet ends with black polyethylene to exclude light from entering. Tie a nylon mesh screen over the end of the return line to collect any debris returning from the growing channels. Drill holes in the reservoir lid for the entrance of the return pipe and for the inlet line and bypass line. You may also install a 100 mesh filter in the inlet line above the pump before the bypass to collect any extraneous particulate matter.

Remember to make or purchase gullies with the correct size and spacing of plant site holes for the specific crops you wish to grow, either low-profile or vine crops. If you want to grow vine crops, you would need only two NFT channels, spaced about 2 ft apart. Then V-cordon training the plants to an overhead support wire or hook

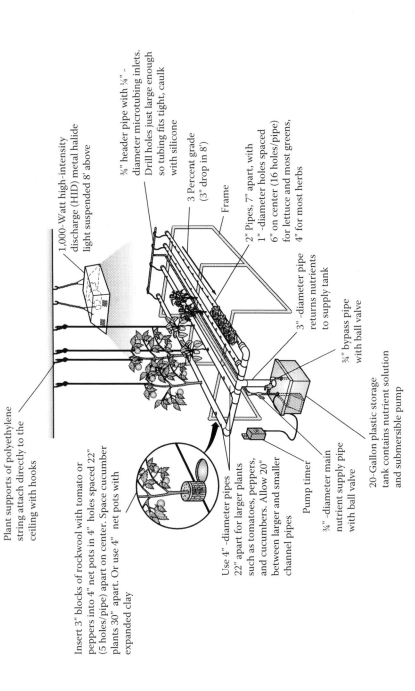

Plant supports of polyethylene string attach directly to the ceiling with hooks

1,000-Watt high-intensity discharge (HID) metal halide light suspended 8' above

¾" header pipe with ¼"-diameter microtubing inlets. Drill holes just large enough so tubing fits tight, caulk with silicone

3 Percent grade (3" drop in 8')

Frame

2" Pipes, 7" apart, with 1"-diameter holes spaced 6" on center (16 holes/pipe) for lettuce and most greens, 4" for most herbs

Insert 3" blocks of rockwool with tomato or peppers into 4" net pots in 4" holes spaced 22" (5 holes/pipe) apart on center. Space cucumber plants 30" apart. Or use 4" net pots with expanded clay

3"-diameter pipe returns nutrients to supply tank

¾" bypass pipe with ball valve

20-Gallon plastic storage tank contains nutrient solution and submersible pump

Use 4"-diameter pipes 22" apart for larger plants such as tomatoes, peppers, and cucumbers. Allow 20" between larger and smaller channel pipes

Pump timer

¾"-diameter main nutrient supply pipe with ball valve

FIGURE 13.12 Frame support and growing channels of small NFT system. (Drawing courtesy of George Barile, Accurate Art, Inc., Holbrook, New York.)

using plastic string and vine clamps as is explained later in Chapter 24 on the training of vegetable crops.

For growing both vine crops and low-profile plants such as lettuce, herbs, and arugula, space the 2″ diameter pipes for the low-profile plants at 7″ centers and then the next two 4″ pipes at 20″ centers. It would be better to use large net pots with some expanded clay substrate for the vine crops to increase root aeration than simply setting the plants in their growing cubes into the bottom of the channels.

WALL NFT GARDENS

In this system arrange the NFT channels mounted with brackets to fences, walls, garages, or any vertical surface having good light (Figure 13.13). Use the standard 4″ wide gullies. The most appropriate crops include lettuce, arugula, basil, strawberries, and herbs. The system consists of two or three gullies of length up to 12 ft, the choice depending upon the available space. A 30-gallon solution reservoir with a lid collects and recycles the solution through an inlet header and return plumbing.

FIGURE 13.13 Wall garden NFT system. (Courtesy of American Hydroponics, Arcata, California.)

MATERIALS

1. NFT channels 8–12 ft long.
2. Wall bracket support system. These are available in hardware and building supply stores. These should be heavy duty forms that screw into the wall. Most have adjustable shelf positions.
3. A 30-gal plastic solution tank with cover.
4. PVC piping to fit the channel drain outlets.
5. Submersible pump, fittings, ball valve, and ¾″ PVC or ½″ black poly inlet hose.

ASSEMBLY

Attach the wall support bracket with 2″ screws to the structure holding the NFT garden. At least two of the vertical bracket plates are required for 6–8 ft NFT gullies and three to four for anything longer. Position the support brackets into the backing plates so that at least a 3% slope is obtained. In this system, the solution will run from one gully into the next one and finally into the nutrient reservoir at the bottom. As shown in the diagram (Figure 13.14), slope one gully in one direction and then the next one in the opposite direction working down so that the solution flows from the lower end of the channel above to the higher (inlet) of the next one below. It is easiest to use flexible hose, like black poly hose, from the drain outlet pipe of the channel above to enter the top of the inlet end of the next channel. The lowest gully has a drain hose into the nutrient tank. Refer to the diagram of Figure 13.14 for details.

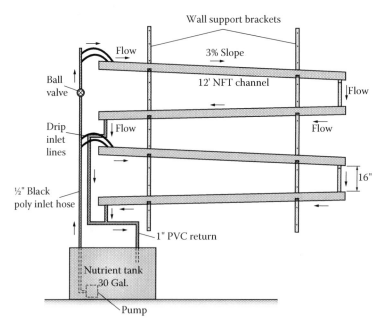

FIGURE 13.14 Wall garden NFT with two sections. (Drawing courtesy of George Barile, Accurate Art, Inc., Holbrook, New York.)

The pump lifts the solution through a ½″ diameter black poly hose from the reservoir to the inlet end of the highest grow channel. Do not use more than two to three channels hooked together or the solution temperature may rise and oxygen deficit may occur to the plant roots. Connecting three 12-ft gutters together in one section is an NFT solution flow through 36 ft. That is very long for growing under high light intensity summer weather. It would be better to use two sections with two channels each for a total of four, providing the height of the wall is sufficient. Each channel must be a minimum of 16″ apart to give adequate light to the one immediately below.

If constructing two separate sections of NFT, use the same inlet header, but install a ball valve before the inlet hose enters the upper gully. This will balance the flow to the two sections. Collect the drainage from the second channel (lower one) and direct the return pipe from each section back to the nutrient tank as shown in Figure 13.14. In effect there are two separate systems but one common inlet header with two individually regulated outlets.

A-FRAME NFT SYSTEM

This design again is suited only to low-profile plants. The concept is the same as that for the wall garden NFT systems, but instead of attaching the growing channels to a vertical surface, they are supported on an A-frame. The length of the A-frame should be 11½ ft to support 12-ft long channels. The size of the A-frame can be varied to fit any length of NFT gullies. I recommend a minimum of 8–10 ft to justify the amount of work needed in constructing the A-frame. Make the framing 6″ shorter than the gullies to enable placement of inlet and return pipes.

Construct the A-frame in the form of an isosceles triangle having two equal sides (Figure 13.15). Locate the first channel 16″ above the floor level so there is enough height to enter the reservoir. Other channels have 10″ between them to reduce any mutual shading by mature plants.

This spacing is sufficient for six channels on each side to give a total of 12 channels on the A-frame as shown in Figures 13.15 and 13.16. The slope of the sides of the A-frame should be shallow enough to permit adjacent channels from overlapping in their horizontal space. Each channel ideally should be offset 4″ horizontally from the adjacent ones next to it. With the sides 75½″ and the base 48″ each channel has a space of 48″/12 = 4″. The altitude of the A-frame is 6 ft. A horizontal bench would contain eight channels in a 4 ft width. The result is an additional four channels or 50% in the number of plants. A 12 ft × 4 ft A-frame with 12 gullies has a total of 12 × 18 plants/gully = 216 plant sites.

MATERIALS

1. The A-frame is to be constructed of ¾″ square or 1″ diameter round galvanized steel tubing. Calculate the length according to the dimensions. The frame members may be welded or bolted. The A-frame may be covered with ¼″ thick white plastic sheeting or a Mylar reflective cover to get more efficient use of light. However, that is not essential.

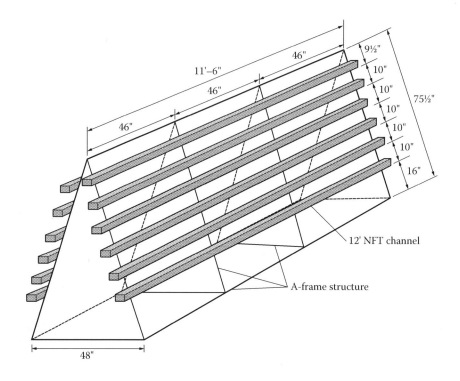

FIGURE 13.15 A-frame with six NFT channels on each side. (Drawing courtesy of George Barile, Accurate Art, Inc., Holbrook, New York.)

2. The reservoir should be 100 gallons for a 12 ft long system or 50–60 gallons for a 6–8 ft long system. Due to the cost and labor in building the A-frame, I recommend to plan on a 12 ft system.

3. A submersible pump with a ¾″ outlet, minimum volume of 10 gpm with a head of 15 ft.

4. One filter of 100 mesh by ¾″ diameter.

5. Inlet piping fittings of ¾″ diameter include male adapters, ball valves, tees, 90° elbows, and ½″ barbed adaptors for ½″ black poly hose and figure "8" caps.

6. Other fittings include drip lines (two per gully), about 30 ft, 24 compensating emitters of 1 gph, and twelve 90° elbows (1″) for drain spouts from the end of the channels to the collection pipe.

7. Twelve 12-ft NFT gullies with covers, end caps, and drain spout. These have holes spaced at 8″ centers. It is best to use channels that have separate covers that can be removed for cleaning.

8. One collection pipe of 2″ diameter—about 10 ft.

9. Collection/return pipe fittings (2″) including caps, 90° elbows, and tees or cross fits with 1″ × 2″ reduced bushings. Follow the diagram plan of Figure 13.16.

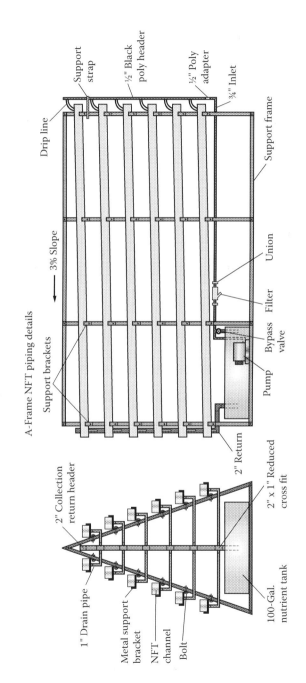

FIGURE 13.16 Details of pump and irrigation system to NFT channels. (Drawing courtesy of George Barile, Accurate Art, Inc., Holbrook, New York.)

10. Support brackets to hold the NFT channels on the A-frame. Use four per channel, so that is a total of 48 brackets. These are bolted to the A-frame bars. Make the brackets from 1″ wide by ⅛″ thick steel. Form their shape to fit the bottom of the NFT gullies similar as shown in the diagram.

Assembly

Construct the A-frame first and later mount the NFT channels onto it after covering the A-frame with sheets of plastic or Mylar, if it is being used. Locate the solution reservoir at the drain end of the A-frame. Install the pump with fittings, a union above the lid of the reservoir and then the bypass and filter. Then, with a tee, make a header to each side of the A-frame under the first NFT gully and then up to the inlet location using a ¾″ × ½″ slip-thread female adapter where the ½″ poly adapter is attached to convert to the ½″ black poly hose. This will be at the base of the first NFT channel on each side. Punch the holes for the emitters; install them and the drip lines to the top of the channels (two per channel).

Mount the 2″ collection pipe header in the middle of the A-frame at the drain ends of the NFT gullies. Use several 90° elbows with a short 4″ piece of pipe at the lower end of the collection pipes to enable making a bend from the angle of the A-frame to the horizontal section joining the two collection pipes just above the nutrient reservoir as shown in Figure 13.16. Use a 90° elbow to enter the nutrient tank. You may put a short spout at the end of the elbow, but do not glue it so that you can remove it when accessing the reservoir for cleaning. Place a cap on the top end of the 2″ collection pipe, but do not glue it in case access to clean the pipe is necessary. Attach the 1″ drain pipes from all of the NFT gutters to the 2″ collection pipe header with tees or a cross fitting. Use only silicone to seal the entrance of this spout into the collection pipe as at times the gullies may have to be removed. Just a note here that when cleaning the gullies after a crop, use a piece of towel soaked in a 10% Clorox solution and scrub the channels. If you have other plants still in the other channels, plug the drain line in the gully with another towel before using the bleach solution to prevent the bleach solution from going back to the nutrient tank. After cleaning, dry the gully with a towel and let the channels air dry before placing in your seedlings.

Drill holes in the nutrient reservoir lid to accommodate the pipe from the pump, the bypass line, and the return drain line.

If you want to place a covering over the A-frame do that now before securing the support brackets for the NFT channels. Remember to locate the support brackets to give a 3% slope to the gullies from the inlet end to the outlet end, that is, about 4″ of slope. The support brackets must be screwed, using self-drilling metal screws or bolts, into the vertical members of the A-frame. Secure the brackets at the inlet and outlet ends first, then attach those on the other frames to get their position accurately.

Position the NFT channels, make the inlet and drain connections, and put water in the reservoir to test the piping joints. Balance the ball valve at the bypass to get the correct flow into the gullies.

This is the most complex and difficult system to build, so do not be discouraged. There are many other more simple and equally productive hydroponic systems to construct as alternatives to the A-frame NFT system.

EBB AND FLOW SYSTEM

In a larger hobby ebb and flow hydroponic system use a series of trays sitting on a framework above a nutrient reservoir. This is a recycle system where the solution floods the substrate from below and then drains back to the reservoir awaiting the next irrigation cycle. A time clock or controller activates the pump several times a day to flood the growing beds. The frequency of irrigation cycles, as mentioned in the previous chapter, is dependent upon the crop, stage of plant growth, and substrate water retention.

The ebb and flow method is suitable to most crops including low-profile and vine crops. Since the substrate of choice must have good porosity, use some form of aggregate. In this case, we could also call it gravel culture. Use expanded clay, ¾″ crushed igneous rock, or ¼″ pea gravel.

In this system beds are constructed 2 ft wide by any length up to 20 ft by 10″ deep. Construct one or two beds with a 3 ft aisle between them for access. Make the beds in the same way as was done earlier for the raft culture raceway of the narrower 2 ft width as shown in Figures 13.2 and 13.3. Build the sides of the bed with 2″ × 10″ treated and/or painted cedar lumber. The bottom is ¾″ plywood. All lumber joints are screwed and glued. The beds are lined with a 20-mil vinyl as for the raceways using the same methods. Each bed must be supported on a wooden or steel framework.

MATERIALS

1. Lumber of 2″ × 10″ dimension for all sides and ¾″ plywood for the bottom of the bed. Since plywood comes in 4 ft × 8 ft sheets, make the width of the bed(s) 2 ft outside dimensions to achieve optimum use of the plywood.
2. Swimming pool 20-mil thick vinyl.
3. Vinyl cement.
4. PVC pipe and fittings.
5. Lathes 1″ × ¼″ thick.
6. Submersible pump with a timer.
7. White oil-based paint and primer.
8. Nutrient reservoir. The volume depends upon the total volume of void spaces in the substrate of the bed(s).

   Calculate the total volume of the bed and multiply it by the percent of void spaces in the rock substrate. Typical crushed ¾″ rock has a void space of about 38%. Pea gravel would be less, closer to 25%. The finer the material the less void space is present. If you wish to test your substrate for its void space, place a given volume of gravel in a container such as a 5-gal bucket and add water to it until the water level just reaches the surface. Pour off the water into another bucket and measure this volume of water with a graduated cylinder. The fraction of that volume of water over the volume of the gravel gives you the percent of void spaces. To allow for some loss of water by the plants make the nutrient reservoir large enough to contain at least twice the void space of the substrate.

9. Metal frame to support the aggregate ebb and flow beds. This should be constructed of 1″ square tubing as the weight will be substantial, especially during an irrigation cycle when the bed is full of solution.
10. Use light weight expanded clay aggregate (LECA).

Volume Calculation

Here is an example to calculate the volume on aggregate needed: Volume = length × width × height (V = LWH). For a 10 ft long bed by 2 ft wide by 9″ deep: V = 10 ft × 2 ft × $\frac{9}{12}$ ft = 15 cubic ft. One cubic yard is equal to 27 cubic ft, therefore the number of cubic yards is: $\frac{15}{27}$ = 0.56. This is slightly over one-half a cubic yard. Order one cubic yard and there will extra for use later.

Weight Calculation

The LECA weighs 1200–1300 lbs/cu yd, whereas, ¾″ crushed gravel weighs about 2800 lbs/cu yd. The total weight for a 10 ft × 2 ft bed of LECA aggregate is therefore: 0.56 × 1200 lb = 670 lbs. Obviously, you need to make the bed and the supporting frame strong enough to withstand this weight.

Solution Volume Calculation (Using a 10 ft Bed with Expanded Clay)

V = 15 cu ft × 35% void space = 5.25 cu ft. Multiply by two for evaporation loss by plants: 5.25 × 2 = 10.5 cu ft. Convert to gallons: 1 cu ft = 7.48 U.S. gal; therefore: 10.5 × 7.48 = 78.5 gal. In this case, use a 100-gal nutrient reservoir to give adequate solution.

Assembly

First construct the bed(s) using 2″ × 10″ lumber for the sides and ¾″ plywood for the bottom. Be sure to use screws and glue in the joints. Install the vinyl liner as described earlier for the raft system raceway, but the plumbing will differ. Install a 1½″ diameter inlet pipe in the center of the bed in the middle lengthwise using bulkhead fittings to seal the pipe with the vinyl liner as discussed earlier for the raceway. An overflow pipe of the same size diameter and 8″ high is placed in the bed with similar bulk-head fittings within 12″ of the end of the bed nearest the nutrient reservoir. This maintains the solution level in the bed during an irrigation cycle.

Make the support frame next before the irrigation system from the pump and the drain line back to the reservoir. Using galvanized or chrome steel square tubing is preferred over wood for the bed framework. Be sure to make a separate frame for each bed. The top of the framework should be 30″ high to give sufficient height above the nutrient tank. The width should be about 1″ wider than that of the bed. Due to the weight of the aggregate and solution make cross supports every 3–4 ft. Tie the entire framework together at the base and top with bars both across and lengthwise as shown in Figure 13.3. At the edges of the top cross bars where the bed sits, extend the vertical tubes 1″ above to act as a guide to contain the bed.

After placing the bed on top of the framework start the plumbing. A submersible pump with a 1½″ outlet and capacity to fill the bed within 5 min will ensure rapid fill and drain of the bed. The volume of the void spaces in the bed in our example of a

10 ft long bed was 5.25 cubic ft or about 40 gal. The pump must be capable of filling the bed with 40 gal within 5 min so the pump outlet volume should be at least 8 gpm (40 gal/5 min). Assemble the piping from the pump in the following sequence: male adapter and bushing to 1½″ diameter pipe; 1½″ union; a bypass pipe with a 1½″ × 1″ tee (or 1½″ tee plus a 1½″ × 1″ reduced bushing); 1″ ball valve on a 1″ bypass line; 90° elbows to take the 1½″ inlet line from above the solution reservoir underneath the supporting framework for the bed and then entering the center of the bed. Within 2″ of the inlet to the bed install a 1½″ union so that the line can be dismantled if necessary. The return overflow line should be 1½″ diameter from one end of the bed closest to the nutrient reservoir. This also can be fastened to the underside of the bed framework. Drill holes in the cover of the nutrient tank for the bypass, main inlet, and return pipes. The piping differs somewhat from the raceway of Figures 13.3 and 13.4 in that an air pump is not needed and the ¾″ inlet line is replaced with the 1½″ main to the bed.

Fill the bed with the expanded clay aggregate to within ½″ of the top. Place a screen over the inlet and overflow pipes before filling with the aggregate. One day before transplanting, sterilize the aggregate with a Zerotol solution (hydrogen dioxide) of 1:100 concentration. This is equivalent to 1¼ fluid ounces per gallon. Water the medium with the Zerotol solution using a watering can from above. Alternatively, put 1 gal of Zerotol in the 100-gal solution tank and pump the solution into the bed several times keeping the pump on for half an hour each time to allow the solution to circulate through the substrate and overflow back to the reservoir. This process will kill most fungi.

ALTERNATIVE SYSTEM

There is one commercial ebb and flow system that uses a series of 5-gal pots, instead of beds, filled with lightweight aggregate. This system is described in detail with drawings (Figures 20.3 and 20.4) in Chapter 20. There is a 6-pot and a 12-pot system. The key components to the system are a 55-gal drum nutrient reservoir; a 5-gal controller bucket; 6 or 12 growing pots with felt liners to prevent debris entering the drain lines; two submersible pumps, fittings, and tubing; two timers; and a float valve.

A 400-gph pump is positioned in the solution reservoir that circulates solution to the controller bucket upon an irrigation cycle governed by one timer. The solution flows from the controller bucket by gravity to each of the grow buckets until they all reach a set fill level that is equivalent in all of the buckets and control bucket. A float valve in the control bucket then stops additional inflow of solution to the control bucket. When the timer for the main pump stops the irrigation cycle to the controller bucket, the second timer activates a smaller submersible pump of 160 gph in the control bucket to start pumping the solution back to the main reservoir. The two timers must be synchronized so that when one starts the other must be off. The solution from the grow pots drains back to the control bucket as the level in the control bucket is lowered by the pump sending the solution back up to the main 55-gal nutrient drum.

One timer is designated the "fill" timer and operates about 20 min during an irrigation cycle. The irrigation cycles are usually ever few hours according to the plant stage of growth and nature of the crop. The other timer called the "drain" timer activates the control bucket pump for about 40 min during a drain cycle as it takes more time for the solution to drain back from the grow buckets than it does to fill them.

This ebb and flow commercial hydroponic system costs about $450. For more information visit the website of "HTG Supply" listed in the Appendix.

You could construct this system yourself. The most challenging aspects would be to locate all of the various fittings needed to connect the distribution hoses, drain screens, and so on to the grow buckets. Each of the pots must be connected to the control bucket with special fittings (grommets) that will seal them to the buckets without leakage. Most are ½″ and ¾″ diameter poly hose fittings.

DRIP IRRIGATION SYSTEMS

Drip irrigation is central to all other hydroponic cultures with the exception of aeroponics. It is used with sand, perlite, vermiculite, small expanded clay, sawdust, rockwool, peatlite mixes, coco coir, rice hulls, and any combination of these as a mixture. Irrigation by a drip system is the common method of providing nutrient solution to the base of plants in all of these hydroponic cultures. Sand and sawdust cultures are contained in beds, whereas peatlite mixes, small expanded clay, and rice hull mixes are better in pots. Rockwool and coco coir cultures use plastic "slab" sleeves. Perlite may use slabs or special pots such as "bato buckets." All use drip irrigation. In the following example, re-cycled systems are used for one bench with bato buckets and the other with high-density rockwool slabs. The larger high-density slabs will support six tomato plants each providing they are V-cordon trained as explained in Chapter 24.

The following components make up a drip irrigation system. The materials listed are for an indoor drip system irrigating two 10-ft rows of pots with two plants each or 8″ wide slabs with six plants each. If these are vine crops, each plant requires 4 sq ft of floor space. Therefore, a growing room of dimensions 12 ft long by 12 ft wide has a total area of 144 sq ft. The maximum number of vine crops in that area is 144/4 = 36 total plants of tomatoes, peppers, or eggplants. European cucumbers must have a minimum of 9 sq ft so the area would contain only 16 plants. You will, however, grow a combination of these crops, for example, 4 cucumbers, 18 tomatoes, 6 peppers, and 4 eggplants. Space the rows 6 ft apart. The first row is 3 ft from the wall and there is 6 ft between it and the next one, making it also 3 ft from the other wall. In our example, (Figure 13.17), the first row with the bato bucket system has 6 peppers, 4 eggplants, and 4 cucumbers and row two in the rockwool slabs has 18 tomato plants.

MATERIALS

1. A submersible or centrifugal self-priming pump with timer.
2. A 35-gal tank with cover.
3. Compensating emitters of 0.5 gph.
4. Drip line of 0.160–0.220″ diameter (36 plant drip lines × 18″ = 54 ft).
5. Barbed stakes: 36
6. Two figure "8" end stops or 3″ × 1″ PVC.
7. Schedule 40, ¾″ PVC pipe: 10 ft
8. Various ¾″ PVC fittings including male adapters, tees, 90° elbows, ball valves, ¾″ × ½″ slip-thread reduced bushings (2).
9. 20 ft of ½″ black poly hose with ½″ barbed adapters (2), 1″ hose clamps (2).

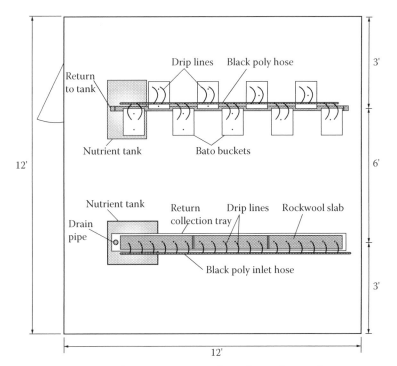

Note: Systems are supported on a frame above tanks.

FIGURE 13.17 Drip irrigation system plan. (Drawing courtesy of George Barile, Accurate Art, Inc., Holbrook, New York.)

10. PVC glue and cleaner.
11. A poly punch tool to make the holes for the emitters in the poly hose.
12. 100 mesh filter.
13. Lumber or ¾″ steel square tubing for supporting the pots on frames or slabs in trays.
14. Lumber to construct trays for slabs.

ASSEMBLY

Here only the assembly of the drip irrigation system is discussed as the remaining growing systems explain the various substrates in pots or slabs and any supporting structure. Starting from the pump, attach a ¾″ main inlet line with a male adapter, then above the tank make a bypass line using a tee, ball valve, and 90° elbow. Above the bypass put a 100-mesh filter in the main line then continue with 90° elbows and the line to a header at the front of the plant rows.

Then use two elbows and a reduced slip-thread busing in each to adapt to the ½″ poly adapters. The ½″ black poly hose is attached to each adapter and runs to the end of each row where a figure "8" end stop is placed. Punch holes for the emitters at the location of the plants and insert an emitter in each. Attach a drip line to each emitter at one end and a barbed stake at the other that sits at the base of each plant.

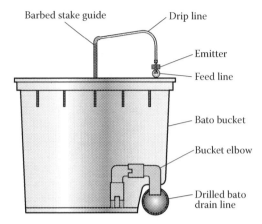

FIGURE 13.18   Bato bucket with drainage siphon at bottom. (Drawing courtesy of George Barile, Accurate Art, Inc., Holbrook, New York.)

PERLITE BATO BUCKET SYSTEM

Bato buckets are special pots designed for using coarse substrates such as perlite and expanded clay. I do not recommend them for finer substrates since the buckets retain a small reserve of solution at the bottom of the pot. Also, any recirculation of the leachate with a fine medium would be more complex in managing due to potential salt build-up. The buckets have an indentation at the back to enable them to sit on a 1½″ drain pipe. A ¾″ diameter double elbow forms a siphon to drain from the pot to the drain pipe (Figure 13.18). This siphon keeps about 1″ of solution in the bottom of the bucket. This persistent reservoir of solution in the bucket is important with coarse substrates.

They are made by a company in Holland and hence are also referred to as "Dutch" bato buckets. These buckets are suitable for vine crops. They are made of rigid plastic, measuring 12″ × 10″ × 9″ deep (Figure 13.18). The bato buckets are placed on a 1½″ or 2″ PVC drain pipe to enable recirculation of the nutrient solution. The rows are spaced 6 ft apart and the bato buckets are staggered at 14–16″ centers within the rows. Some hobby units are available as a self-contained supporting structure, pots, and irrigation system in compact configuration (Figure 13.19). As long as the plants are V-cordon trained to optimum spacing at the top of the crop, the narrow spacing of the pots is functional. Two plants are grown in the buckets, with the exception of one for European cucumbers to obtain the correct growing area per plant. Irrigation is by a drip system.

In an indoor bato bucket system, the pots will have to be raised above a solution tank with a framework of steel or wood. The frame has to be just high enough to permit gravity flow of the recycled solution to the tank as shown in the diagram of a commercially available hobby system (Figure 13.20).

MATERIALS

The following materials list is to build a system of two 10-ft rows of bato buckets with the same irrigation system as described previously under "Drip Irrigation

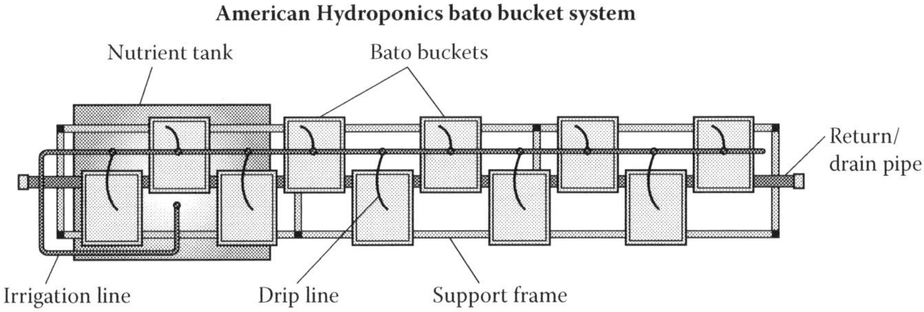

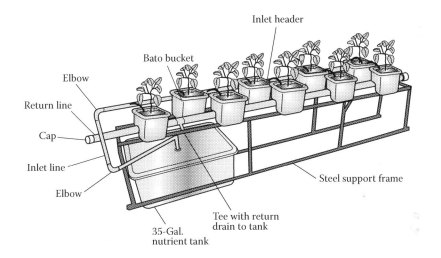

FIGURE 13.19 Bato bucket layout (plain view). (Drawing courtesy of George Barile, Accurate Art, Inc., Holbrook, New York.)

FIGURE 13.20 Side view of supporting frame and layout of a bato bucket. (Drawing courtesy of George Barile, Accurate Art, Inc., Holbrook, New York.)

System." The irrigation system is exactly as outlined; therefore the materials for that portion include all items 1–13 listed earlier with a 50-gal tank. The additional materials needed are listed in the following text:

1. There are nine bato buckets per row, so a total of 18 bato buckets.
2. Thirty feet of PVC pipe of 1½″ diameter for the drain collection lines.
3. Various 1½″ PVC fittings for the drain return to the solution tank include 90° elbows (3), female adapter plus threaded plugs (2), and one tee.

ASSEMBLY

Construct the supporting framework as shown in Figures 13.19 and 13.20. Make two individual benches, one for each row of pots. You may place ½″ thick plywood on

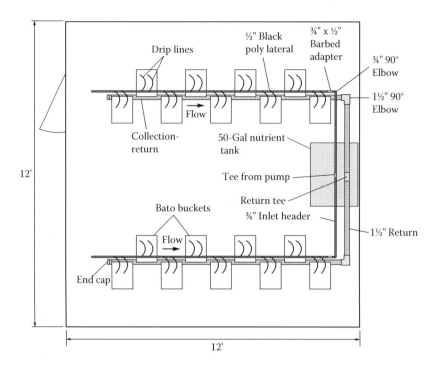

FIGURE 13.21   Details of drainage and irrigation systems for two tables of bato buckets. (Drawing courtesy of George Barile, Accurate Art, Inc., Holbrook, New York.)

the top of the benches, but it is not necessary for bato buckets as they are supported at their drain end by the drain pipe. Once the framework is completed make up the drip irrigation system as described earlier. Then, secure the drain pipe system on the frame under the drain end of the bato buckets. Cut 1″ diameter holes along the top of the drain pipe at the positions of the pots (14″ centers starting 8″ in for the first one). Join the two drain pipes with elbows and a tee to a common header that then enters the top of the solution tank. Locate the nutrient tank between the rows at the back wall. Connect all of the inlet pipes, bypass, filter to the pump and make up the header with risers up to the top of the first bato bucket in each row and connect, with an elbow and adapter, the black poly hose with the emitters and drip lines as shown in Figure 13.21.

One day prior to placing the pots with the perlite, moisten the perlite and flush the substrate with Zerotol (1¼ fl oz per gal). Upon positioning the bato buckets, place the black poly hose along the top of the pots. The pots are now ready to transplant. During transplanting put one drip line with a barbed stake at the base of each plant. Irrigate 4–5 times per day with 5 min duration per cycle. Adjust the frequency of cycles with the stage of plant growth and cycle duration for 20% leachate.

ROCKWOOL CULTURE

Start seedlings in rockwool cubes that are transplanted to rockwool blocks before a second transplant (about 5–6 weeks after sowing for most vine crops except cucumbers about 2 weeks) to rockwool slabs. Rockwool properties and products were discussed in Chapter 11.

For consistency in description of materials and components, the same size of rockwool system as for perlite bato buckets is presented that grows 36 vine crops in an area of 12 ft × 12 ft. There are two rows of plants, but instead of bato buckets, use 8″ wide rockwool slabs. The supporting framework for each bed differs somewhat in that it should slope 3% toward the solution tank.

A tray under the rockwool slabs collects the leachate from the rockwool slabs and returns it to the nutrient reservoir. To contain 8″ wide slabs construct watertight trays 10″ wide by 4″ high by 10 ft. Each tray has three slabs each with six plants of tomatoes, peppers, or eggplants as shown in Figure 13.17. Grow only three European cucumbers per slab to meet spacing demands of the plants. V-cordon train the plants as outlined in Chapter 24. To make the return tray waterproof, line it with 20 mil vinyl, folding and gluing it as was shown in the raceway construction in raft culture. A 1½″ diameter drain pipe is installed at the lower end within 2″ of the tray end in the same way as described for the raceway. The drains of the two trays are attached with a union to the collection pipe header going to the solution tank. The irrigation system is the same design as for the bato bucket perlite system shown in Figure 13.21. Items 1 through 13 of the materials for the drip irrigation system with a 50-gal tank are used in addition to the following.

Materials

1. Six rockwool slabs 4″ thick by 8″ wide by 39″ long.
2. Fifteen feet of PVC pipe of 1½″ diameter for the drain collection lines.
3. Various 1½″ PVC fittings for the drain return to the solution tank include 90° elbows (5), unions (2), bulk-head fittings (2), and one tee.
4. 1″ Styrofoam to insulate the slabs.
5. ¾″ thick plywood.
6. Fifty feet of 1″ × 4″ lumber for tray sides and ends.
7. Two pieces of 20-mil vinyl swimming pool liner 2 ft wide by 12 ft.
8. Vinyl cement.
9. White oil-based paint.

Assembly

The benches for the slab trays are constructed the same as for the bato buckets with the exception of the 3% slope to the solution tank. The inner dimensions of the trays are 10″ × 10 ft × 3½″ high. Cut the plywood to 10 ft 4″ long by 11½″ wide. This is the bottom of the tray. Screw and glue the 1″ × 4″ boards around the perimeter of the plywood. Paint the wood after assembly, but before placing the vinyl liner. Drill the drain hole within 2″ of the end of the tray to get a snug fit for the 1½″ diameter pipe and seal it with bulk-head fittings as described in the construction of raceways earlier and as shown in Figure 13.22.

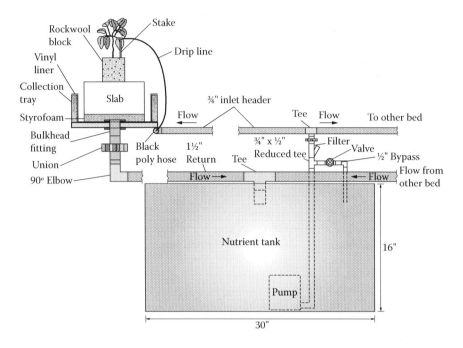

FIGURE 13.22   Rockwool culture layout. (Drawing courtesy of George Barile, Accurate Art, Inc., Holbrook, New York.)

Once the collection trays are completed, put them in place and make up all of the return lines and attach the header to the trays with a union so that they can be dismantled for cleaning and/or repair. Then, fabricate the drip irrigation lines with the inlet pipe, bypass, and so on from the pump.

Test all of the irrigation and drainage systems with raw water in the solution tank to detect any leaks.

A key component is the 1″ thick Styrofoam. Use the high-density "Roofmate" type of Styrofoam. Cut the Styrofoam strips 8″ wide and a total length of 9 ft 8″ long so that they do not cover the drain as shown in Figure 13.22. The Styrofoam strips are placed immediately underneath the slabs. The Styrofoam between the bottom of the return tray and the slabs has two important functions. First, it insulates the slab keeping the roots warm. Second, it raises the slabs up slightly and away from the sides of the tray to permit easy flow of the leachate. After the slabs have been presoaked for several hours, cut drainage slits 1″ long on a 45-degree angle from the base of the slab liner at the level of the Styrofoam underneath. These drainage cuts are positioned between the plants. Three cuts are sufficient. Cut them on the same side (only one side) of each slab.

Seedlings are sown in 1½″ rockwool cubes, transplanted to 3″ blocks, and finally to the slabs after 5–6 weeks from sowing. Remember to presoak the slabs with a half strength nutrient solution prior to transplanting the seedlings in the blocks. At that time make up the nutrient solution in the tank.

Frequency of irrigation cycles is dependent upon stage of plant growth and the crop. Make the duration of any given cycle long enough to get at least 20% leachate.

COCO COIR CULTURE

The entire setup for coco coir culture is the same as for rockwool culture. The slabs for coco coir are the same dimensions as those for rockwool; however, they also make "mini" slabs that are shorter than the normal ones and are good for two plants only.

The only difference between these two cultures is the frequency and duration of irrigation cycles. Coco coir retains much more water than rockwool, so fewer irrigation cycles are needed. Leachate can be reduced to about 15% maximum with a shorter duration of any cycle.

Seedlings may be started in coco coir cubes and blocks such as those sold by Jiffy Products, but most growers still use the rockwool cubes and blocks and transplant to the coco coir slabs by setting them on top of the slabs as is done with rockwool slabs.

AUTOPOT SYSTEM

This is basically a wick type of system of a series of pots, but a patent system due to a special "AQUAvalve" that regulates the irrigation to the pots. For details visit their website (see Appendix) and Figures 13.23 and 13.24. They recommend using substrate 50/50 mixes such as coco/perlite, coco/expanded clay, and rockwool/expanded clay.

The AutoPot requires no power, pumps, or timers to operate. It functions on the gravity of solution from a tank. This flow of solution to the pots is regulated by the AQUAvalve through gravity from the solution tank. It fills a tray under two pots to about ¾" in depth and does not fill it again until the solution is used up by the plants in the pots. Each tray holds two pots. A capillary-like disc is set in the tray under the pots. A black "matrix" disc is set in the bottom of the pots before filling with a substrate. This also acts in bringing the solution in contact uniformly with the substrate.

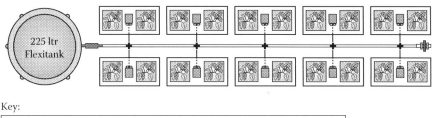

Key:

—— 6 mm pipe	⌁ 16 mm inline tap
▭ 16 mm pipe	▱ 16 mm click-frt tank adaptor and filter
✦ 16–6 mm cross connector	▨ AQUA valve

FIGURE 13.23 AutoPot system layout with 20 pots. (Drawing courtesy of George Barile, Accurate Art, Inc., Holbrook, New York.)

FIGURE 13.24   AutoPot system of 20 pots. (Courtesy of AutoPot Global, Ltd., Paddington, Oxfordshire, United Kingdom.)

AEROPONICS

This culture is not feasible on a larger scale than the small indoor units that were described in Chapter 12. A somewhat larger unit may be constructed as an A-frame. In this system make an A-frame of 1¼″ diameter PVC pipe and secure 1″ thick Styrofoam on the outside to hold the plants. The frame must sit above a nutrient tank. The width of the base should be a few inches narrower than the nutrient reservoir so that the solution will drain back to the reservoir. The system can grow low profile crops and medicinal plants whose roots are harvested for vitamins or drugs.

The following list of materials is for an A-frame of 3 ft wide base and 4 ft long. The height of the sides is 4 ft. The sides are therefore 4 ft × 4 ft with the bottom edge sitting on top of the nutrient reservoir that is 3 ft × 4 ft × 9″ deep. Construct the nutrient reservoir of wood and line it with 20-mil vinyl as was done for the raceway system.

MATERIALS

1. Lumber for the reservoir includes 14 ft of 2″ × 10″ and ¾″ thick plywood for the bottom.
2. A piece of 20-mil vinyl liner 6 ft × 8 ft.
3. A-frame materials include: 1¼″ diameter schedule 40 PVC pipe, 90° elbows, and tees.

4. A high pressure submersible pump with 200-mesh filter.
5. PVC pipe of ¾″ diameter schedule 80 supporting and connecting to the mist nozzles.
6. Various PVC fittings for the mist distribution pipe.
7. A 24-h timer with minute intervals.
8. Two sheets of 4 ft × 8 ft × 1″ thick Roofmate Styrofoam.
9. Various hardware of bolts, plastic electrical ties, screws, glue, and so on.

ASSEMBLY

Construct the nutrient reservoir with outside dimensions of 4 ft × 3 ft. Be sure to use screws and glue on all joints including the plywood bottom. Fold in the vinyl liner as described for the raceways previously. Paint the reservoir with white oil-based paint before lining it. This reservoir has no drain.

Once the reservoir is completed build the A-frame to fit onto it. Assuming that the exact outside dimensions are 48″ × 36″, make the A-frame slightly smaller to allow for the 1″ Styrofoam that will cover the sides and ends of the A-frame. The inside of the Styrofoam must overlap the reservoir frame by ½″ to permit the moisture when irrigating to run back into it. Refer to the drawings of Figure 13.25 for details. The outside dimensions of the A-frame without the Styrofoam then is as follows: width: 33″ less ½″ on each side = 32″; length: 48″ less ½″ on each end = 47″. First, construct the two ends using 1¼″ PVC pipe as shown in Figure 13.25. The lower base cross member is 36″ long in order to rest on top of the nutrient reservoir sides. One horizontal 1¼″ pipe (47″) going lengthwise is attached by stainless steel screws or bolts in the corners of each "A" to complete the frame.

Fit the two 36″ cross members of 1¼″ diameter PVC with a 90° elbow on each end at exactly the outside width of the nutrient reservoir so that these supports will fit snugly against the outside wall of the reservoir to strengthen it. These are attached to the top sides of the tank using 2″ self-tapping stainless steel screws. Fasten with self-trapping stainless steel screws the peak of the frames together using the horizontal member at the top angle.

Cut the Styrofoam to fit the sides and ends securing them to the A-frame with stainless steel self-tapping screws. Be careful that your measurements are correct so that the insides overlap into the solution reservoir about ½″ to allow drainage back to the solution tank to prevent leakage outside. Refer to the diagrams of Figure 13.25 for details. Cut ¾″ diameter holes in the side Styrofoam boards at 6″ × 6″ for lettuce, arugula, basil, and some herbs or 4″ × 4″ centers for smaller herbs. This is similar to the spacing for boards of raft culture so the 4 ft × 4 ft board would hold 64 plants at 6″ × 6″ spacing. Leave the Styrofoam ends of the A-frame off until the irrigation system is completed.

Make up the irrigation system using ¾″ diameter Schedule 40 PVC from the pump. Install the 200-mesh filter at the front of the header. Place the header with the mist nozzles on top of 1″ PVC frame cross members at 12″ above the nutrient reservoir. Install four mist fogger heads at 12″ centers in line along the top of the pipe. The position of the first mister is 6″ from the end and the other three are at 12″

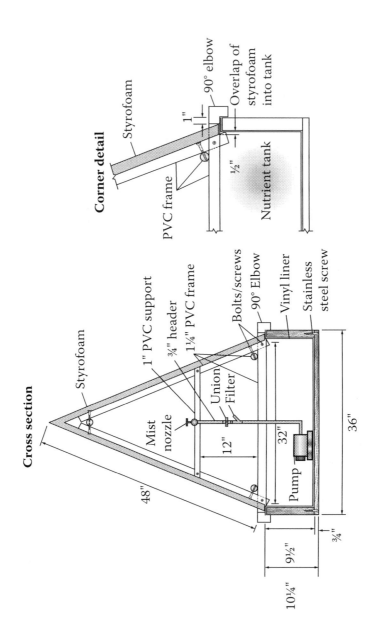

FIGURE 13.25 Aeroponic A-frame cross section with details. (Drawing courtesy of George Barile, Accurate Art, Inc., Holbrook, New York.)

centers making the fourth one 6″ from the opposite end. A timer activates the pump to operate the mist cycles. The frequency and duration of the cycles depends upon the crop and its stage of growth, but as a guideline fog every few minutes for 5 sec, day and night.

PLANT TOWERS

This is a system developed to increase the number of plants in a unit area occupied. This system is for low-profile plants and is particularly suitable to most herbs and strawberries. Plant towers need about 10–12 sq ft of floor area to allow for adequate light. The best method of construction is with the use of special Styrofoam pots available in the marketplace (see Appendix for suppliers). While in some cases you can grow bush tomatoes, peppers, or eggplants, light must be sufficient to penetrate the whole crop canopy. With those crops put only half the number of plants in a tower. With the square pots that are available, plant in the corners of the pots. The plant towers must be supported with a central pipe and be secured above by a cable or other bracket to the ceiling. The towers use a drip irrigation system, two drip tubes in the top pot and one in the middle pot of the tower. The towers must sit on top of a collection pots connected to piping that will conduct the solution back to a nutrient reservoir. I do not recommend using polyethylene sacks as they break with age. You can also construct a column from large diameter PVC pipes such as 8″ diameter. But, again the square Styrofoam pots are better to make transplanting or seeding with subsequent plant support easier than will be the case with pipe columns or sacks.

In the previous example for the bato buckets in a 12 × 12 ft growing space, we can fit about 10–12 plant towers. Here is a list of supplies needed for 10 plant towers. Seven pots per tower are sufficient as the pots are 9″ × 9″ × 8″ tall. Seven pots is a height of 56″. In addition to the pots is the height of the collection pot, at least 10″, plus the support frame height of 20″ for a total tower height of 86″. That height will just fit in a normal house ceiling height of 8 ft with some room above for the irrigation system, tower supports, and so on.

The towers are at 24″ centers with the first located 18″ from the wall. There are five plant towers per row and two rows spaced at 3 ft – 6 ft – 3 ft distances starting from the side wall of the room as shown in Figure 13.26. A 2″ diameter PVC collection pipe at the base of the plant towers returns the solution to the nutrient tank.

MATERIALS

1. For up to 10 plant towers use a plastic storage bin of 50 gal as a nutrient tank.
2. If the plant towers are seven pots high you need 70 pots for 10 towers.
3. Thirty feet of 2″ PVC schedule 40 pipe for the drainage return pipe.
4. Fifty feet of drip line.
5. Thirty compensating emitters of 1–2 gph.
6. Time-clock controller with 24-h and 60-min increments.
7. One submersible pump having a lift capacity of 10 ft with 30 psi pressure.

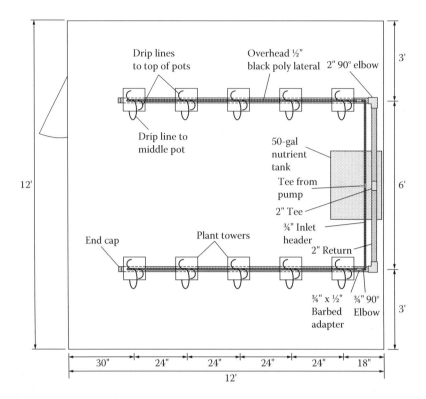

FIGURE 13.26 Layout of plant tower system with all components. (Drawing courtesy of George Barile, Accurate Art, Inc., Holbrook, New York.)

8. Twenty feet of ¾″ PVC Schedule 40 pipe as the main to the height of the plant towers.
9. Thirty feet of ½″ black poly tubing.
10. Various PVC fittings including a ¾″ bypass ball valve, four ¾″ male adapters, six 90° elbows (¾″), two ½″ slip-thread reduced bushing, and two ½″ male barbed adapter to poly tubing.
11. Punch tool for making the holes for the emitters for the black poly lateral tubing.
12. Ten collection pots for the base of the plant towers.
13. One hundred feet of ¾″ diameter galvanized electrical conduit to support the plant towers.
14. Various lumber or ¾″ square steel tubing to build a support stand for the towers to keep them above the level of the nutrient reservoir.
15. One hundred feet of 1″ thin wall PVC for the sleeve over the conduit pipe to permit easy rotation of the plant tower.
16. Ten rotation disks of ¼″ thick by 3″ × 3″ square plastic plate. Drill ¾″ hole in center to permit conduit to pass through it.
17. Bulk-head sealed fitting for each collection pot or tray to connect ½″ black poly line to return pipe.

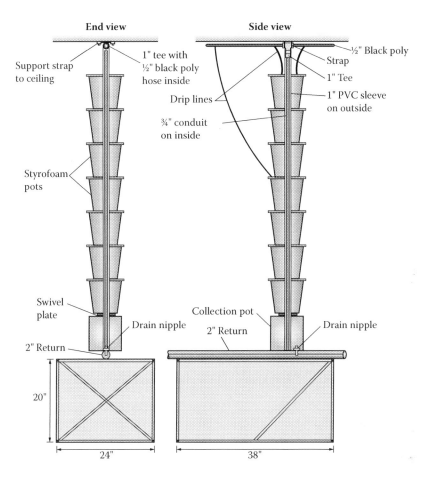

FIGURE 13.27 Irrigation piping from the pump to the top of the plant tower. (Drawing courtesy of George Barile, Accurate Art, Inc., Holbrook, New York.)

ASSEMBLY

First construct a supporting frame 24″ wide by 20″ high by 114″ long for the plant towers with the nutrient reservoir positioned underneath or in the center as shown in Figure 13.26. Attach a vertical member every 38″ along the length. Put braces on the end portions of the frame as shown in Figure 13.3 for the raceway system. The plant tower can be either set on top of the drain pipe as shown in Figure 13.27, or it can be set beside the drain pipe with a small hose line attached from the collection pot to the return pipe. If the plant tower is set beside the return pipe, cover the top of the frame with ¾″ plywood. Placing the collection bucket on top of the return pipe requires a sealed drain nipple from the bucket to enter the return pipe. Place the bottom end of the conduit in the collection pot to the side of the drain nipple after sliding the swivel plate and a 10″ long 1″ diameter PVC sleeve over the conduit as shown in Figure 13.27. This piece of pipe must be the same length as the height of the collection pot or one inch shorter so that the tower will drain into the collection pot. Assemble the plant towers using the steel conduit and the 1″ PVC sleeve. Then, slide the special (Vertigro) Styrofoam pots over the conduit with its sleeve.

Stack the pots so that each is rotated by 45° to the one below and set them in the special locking indentations. Use a maximum of seven pots per tower. Put a 1″ diameter PVC tee at the top of the conduit and fasten the tee to ceiling with pipe strapping or other type of bracket. To permit disassembly of the plant tower remove the bracket holding the tee and slide out the remaining plant tower assembly. Once all of the plant towers have been placed at 2 ft centers, start on the irrigation and drainage system.

From the pump in the solution reservoir attach a ¾″ main using the male adapter, tees, and elbows to a bypass line as was shown in rockwool culture (Figure 13.22). Past the bypass return line with the ball valve, continue the header up to within 2″ of the ceiling where a ¾″ × ½″ slip-thread adapter is connected to a tee at the top. Connect ½″ black poly hose to the ½″ barbed adapters of the tee going both directions to the plant tower rows. Support the PVC tee with a pipe strap or other bracket to the ceiling. The black poly hose attached to the barbed adapters passes through the 1″ tees at the top of each plant tower support pipe and is then plugged at the end using a figure "8" adapter or a short piece of 1″ pipe. Punch three holes in the black poly hose close to each tower and insert the emitters and then the drip lines to them. Two lines go to the top pot and the other one goes to the center pot of each tower.

Attach the bulk-head fitting to the lower side of each collection pan and connect it with a short piece of ½″ black poly hose to the 2″ return pipe to the solution tank. Be sure that the return pipe is lower than the base of the collection pan so that all solution will drain back from the pan to the return pipe by gravity. Alternatively, use a sealed ½″ or ¾″ PVC nipple on the bottom of the collection pot to enter the return pipe located immediately below as shown in Figure 13.27. The return pipe is plumbed back to the top of the solution reservoir as shown in Figure 13.26.

Fill the pots to within 1″ of the base of the adjacent upper pot with substrate of your choice depending upon the crop grown as outlined in the table of Chapter 15. Start herbs by seeding directly into the substrate of the pots. When seeding herbs be sure to use 8–10 seeds per corner planting site. With basil, bok choy, and arugula start the seeds in growing cubes and transplant to the pots after a few weeks. With strawberries you must purchase pre-chilled plants or bring in runners from your garden in the fall or early spring as they must go through a dormancy period. Start up the system to check for any leaks if these plant towers are located in your house as all solution must be contained to prevent damage to your floors.

14 How to Start Your Plants

SEED VIABILITY AND PERCENT GERMINATION

For the growing of hydroponic vegetable crops start your plants from seed. All seeds have a certain life expectancy called viability. The viability is dependent upon the seed itself as well as age and conditions under which it was stored. To lengthen viability keep the seeds in a refrigerator at about 40°F (4.5–5°C) or slightly higher. Do not put them in the freezer section. In general, large seeds retain their viability longer than small seeds. Lettuce, for example, loses viability after 6 months. As viability falls, so does the percentage germination. This is the percent of seed that will germinate under normal sowing conditions of a moistened medium. Every seed package will give percentage germination at the time of packaging the seed. The date tested is given on the seed package. Tomatoes will retain viability for about 3 years when stored in a cool, dry location. However, the percentage germination will decrease over this time period.

You may test the seed germination quite easily. Put 100 seeds on top of two to three layers of moistened paper towels. Then, cover them with another two to three layers of moistened paper towels. Keep them at about 70°F to 75°F (21–24°C) or whatever germination temperature is recommended on the seed package. Inspect the seed after several days, or after the expected length of time indicated on the seed package for the seed to germinate, and count how many of them have broken the seed coat and started to grow. Take that number of germinated seeds and divide by the total of 100 to get the germination percentage.

This percent of expected germination is important for you to determine how many seeds to sow. For example, if 70% of the seeds germinated, then you know you must sow at least 30% more seeds to get the number of plants you want. As an example, if you want 20 tomato seedlings and you found a 70% germination, then multiply the 20 by the percent you want (100%) and divide by the percentage germination (70%): $20 \times 100/70 = 29$. Sow 30 seeds to obtain more seedlings than needed. In that way, you can select the most vigorous seedlings to transplant.

SEED SOURCE AND VARIETY

Another important factor is to get the best seed available. You may order through many seed distributors such as those listed in the Appendix. The next thing to consider is the choice of variety. In most cases, greenhouse varieties will do best with indoor hydroponics. Do not collect seeds from your plants and volunteer them for the following crop as most varieties are F1 hybrids and will not produce the same plant in the next generation. The choice of varieties is discussed in more detail in Chapter 22.

PLANT HABIT/FORM

The form of the plant is also important in your choice. When growing indoors you may use bush varieties, but staking ones are more productive. These are also termed "indeterminate." Bush or "determinate" varieties grow to a certain height and then stop, whereas the staking types continue to grow upward. This applies to vine crops like tomatoes, peppers, eggplants, and European cucumbers and not to low-profile crops.

SEEDLING GROWING SUBSTRATES

When starting seedlings for hydroponic culture we use a soilless medium. The easiest method of sowing seeds is to use growing cubes. There are many available in the market. If you want to grow in peatlite mixture or coco coir in pots, you may use peat pellets or compact cell trays that you fill with a peatlite or vermiculite substrate. For lettuce, arugula, bok choy, basil, and herbs, it is best to start the seeds in either rockwool or Oasis cubes as shown in Figures 11.3 and 23.8. There are a number of different sizes of these products to choose from. Use the smaller ones, usually 1″ × 1″ × 1½″ rockwool cubes or 1″ × 1″ × 1½″ Oasis "Horticubes" for lettuce, arugula, and so on. With vine crops use the 1½″ × 1½″ × 1½″ rockwool cubes.

When growing vine crops, the seedlings in the rockwool or other cubes are transplanted to rockwool blocks before later placing them in the final growing substrate. The advantage of starting the seed in growing cubes is that you may select the best plants later to be transplanted to the rockwool blocks. This is another reason to sow at least 10% more seeds than the number of plants needed. Growing cubes and blocks must be thoroughly saturated and flushed with raw water prior to sowing or transplanting. Rockwool has a pH between 7 and 8.5 and hence is very basic. Reduce it to optimum levels between 6.0 and 6.5 by using a slightly acid solution after saturating the cubes. This can be done by adding vinegar or acetic acid to water to adjust the pH within this optimum range and soaking the cubes again before sowing the seeds. When preparing the blocks for transplanting, soak them with a half strength nutrient solution having a pH between 6.0 and 6.5.

Rockwool blocks come in various sizes: 3″ × 3″ × 2.5″, 3″ × 3″ × 4″, 4″ × 4″ × 2.5″, and 4″ × 4″ × 3″. The choice of size is a function of the plant grown, the size of the vtransplant, and the stage at which you wish to transplant to the final growing area. The longer the plant is held before the final transplant stage, the larger the block. Place the seedlings in the growing cubes or blocks in plastic mesh trays to permit rapid drainage. The trays should sit on a propagation table of wood or galvanized steel with either wire mesh top or lumber with 1½″ spaces between each cross member. This promotes rapid draining and allows aerial pruning of the plant roots that may grow from the cubes or blocks. Do not allow a root mass to form under the cubes or blocks as damage to the roots will occur during transplanting that will predispose the plants to diseases.

PROPAGATION BENCH

If you wish to contain the drain water from the seedlings after irrigation, build an ebb and flow bench that sits on top of a ¾″ steel square tubing framework, as described earlier in Chapter 13, Figures 13.2 and 13.3, and Figure 14.1. However, make the dimensions different from those described in Figures 13.2 and 13.3. Make the inside width 49–50″ by 4 ft 2″ or greater in length, depending upon the number of seedling trays, by 3″ deep. Each seedling tray is 1 ft × 2 ft, so a 4 ft × 4 ft bed would fit eight trays. To get complete drainage away from the trays and allow air pruning of roots, place 1″ of igneous rocks in the base of the bed to support the trays above any drainage water (Figure 14.1). Use rocks of ¾″ diameter. This will give large spaces among them that will dry between irrigations.

Alternatively, make a support frame of 1″ PVC pipe as shown in Figure 14.1. This support frame would keep the mesh trays above the drainage level of the ebb and flow tray. A 2″ high overflow pipe similar to that shown in Figure 13.3 would regulate the maximum solution height at 2″ enabling the solution to moisten the seedlings from below. The overflow pipe, drain pipe, nutrient tank, and plumbing are not shown in Figure 14.1 as they are shown in detail in Figures 13.2 and 13.3. The only difference is that the pump is directly connected to the drain pipe in the bed to fill and drain the bed. Use a 1″ feed/drain pipe from the pump. When the pump shuts off the water will flow back to the nutrient tank below through the pump.

The seedling mesh trays are 12″ × 22″, therefore orient the support frame below so that the trays will be placed perpendicular to the frame. The ebb and flow bed can be 25″ wide by a multiple of 1 ft in length to fit the trays most efficiently. If you need a larger area, make the ebb and flow bench 49″ wide by a multiple of 2 ft (such as 4 ft × 6 ft or 8 ft). Add an extra 1–2″ to the length to give ample room for the seedling trays. The irrigation cycles are automated with a timer operated pump in the nutrient tank below the bench.

During the initial seeding of the cubes, the cubes could be placed immediately on the surface of the ebb and flow bench, but it is better to use the mesh trays as that allows complete drainage. This method would also be fine for low-profile plants that are not transplanted to rockwool blocks. The placement of the layer of rocks or the

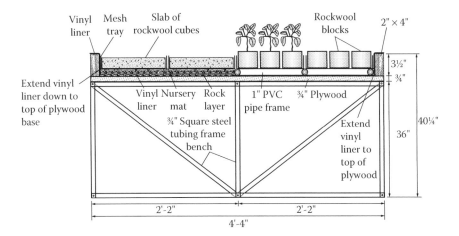

FIGURE 14.1 Propagation ebb and flow bench. (Drawing courtesy of George Barile, Accurate Art, Inc., Holbrook, New York.)

support frame will reduce algae growth in the bottom of the ebb and flow bench due to the exposure to light to a wet surface containing nutrients.

When using the ebb and flow system for the rockwool blocks, keep them in mesh trays and cover the surface of the bed with rock or something else to prevent algae growth. Several layers of black nursery weed matting on top of the rock layer will further discourage algae growth.

Specific details of sowing and transplanting for each crop are given in Chapter 23.

15 Choosing Hydroponic Systems for Specific Crops

The choice of not only the hydroponic system but also the substrate is determined by the crop grown. When making this decision, consider the expected growth of the plant during its cropping period and the form of the crop (low-profile vs. vine-crop) stature. Small plants have less extensive root systems and can hence grow in smaller containers. Many plants simply do not like to grow in a nutrient solution alone but must have a medium into which to spread their roots. These include many herbs such as sage, rosemary, thyme, and oregano that are accustomed to a dry medium around their bases (crowns). Long-term crops that grow for many months cannot tolerate a water solution as they soon develop an oxygen deficit that causes their roots to die.

Plants that can tolerate a water culture system include lettuce, basil, watercress, mint, chives, arugula, and bok choy. Lettuce, arugula, basil, and bok choy have a relatively short cropping cycle of less than 2 months. For this reason, they will grow well in water culture systems such as the floating raft system or nutrient film technique (NFT). Most herbs, if grown on a short cycle, especially for "live" herbs, are happy in water culture systems. "Live" herbs are those that are harvested with their growing cube and roots intact and marketed in small plastic bags or sleeves as "living" plants. This method of harvesting prolongs their shelf life. It is very popular for basil, arugula, and lettuce. Even for your own use, keep the roots on these plants if you are not going to use the entire plant at one preparation. You will be aware of the lettuce in the supermarkets that are packaged in semi-rigid (clam-shell) containers with a lid and a small reservoir in the bottom to contain the roots in their growing cube. The most popular method of growing "live" herbs is in NFT channels, as they can easily be removed at harvesting without damaging a lot of their roots.

Long-term crops growing more than several months must have a substrate and be containerized or grown in beds. These include all vine crops and melons. Rockwool and coco coir slabs are the most common systems of hydroponics with these types of crops. They all use drip irrigation. The advantage of the drip irrigation system is to apply the solution near the base of the plants bringing oxygen with it as it percolates through the substrate. With the moisture homogeneously distributed within the substrate, the roots can spread around to seek water, nutrients, and oxygen. With any open or re-circulating system, it is critical to get good drainage that conducts the leachate away from the plant substrate and container so that oxygenation is not restricted to the plant roots. With ebb and flow systems, this complete drainage between irrigation cycles is one factor that will determine the success or failure of the crop. While small-scale systems of ebb and flow crops will do well as drainage is fairly rapid, with much larger systems of ebb and flow it is more difficult to get adequate and complete drainage in order to avoid restricting plant growth and production. For that reason, with long-term crops I do not recommend using ebb and flow systems. If crops are stressed by environmental limitations, they become more susceptible to pests and diseases. This is another reason to select the correct growing system for the specific crop.

Certain plants have preferences for specific substrates. Some plants like a higher water-holding capacity, while others need very rapid drainage and less moisture contained within the medium. Most vine crops thrive in most media, providing the medium has good drainage with available oxygen to the plants' roots. A few herbs such as rosemary and sage like very good drainage so do best in a coarse well-drained substrate like perlite, vermiculite, or mixtures of these with peat or coco coir at a 1:1 ratio. Knowledge of the type of soil these plants prefer will indicate the nature of medium to use with hydroponics. For example, if the plants do best in a sandy soil, they prefer a substrate with very good drainage, so we would use perlite, vermiculite, or the mixtures of these with some peat or coco coir.

Table 15.1 summarizes the most suitable hydroponic systems and substrates that some commonly grown hydroponic crops prefer. The raft culture system is not suitable to many crops due to the logistics of harvesting. It is used only for one-time harvest plants, so even though some low-profile plants may thrive in the water culture it is not feasible from a labor stand point to grow them as the boards would have to be taken out to harvest and then replaced again with the mature plants intact. This is difficult to do without damaging the boards or the plants. The roots of the plants could become broken or infected with disease organisms during the removing and replacing of the boards into the pond or raceways. Also, with a single harvest crop the boards are then sterilized before placing them back into the raceways for transplanting.

Aeroponics can grow most plants well, but due to the cost of such a system and the inconvenience of supporting vine crops, it is not used for vine crops. It is feasible from a growing point of view for low-profile plants, including strawberries, lettuce, arugula, basil, and many herbs, but again it may not be the correct choice due to capital costs. Aeroponics is the culture of preference for any crop whose harvestable portion is high value roots or tubers. For example, to produce medicinal products from some herb plant roots, aeroponics is both practical and economically feasible. Another very specific application is the growing of seed potatoes as they must be kept free of any disease. This system of aeroponics is used in the International Potato Research Center at Universidad de La Molina in Lima, Peru (Figures 15.1 and 15.2). That center ships certified potato seed tubers to all parts of the world in establishing mother plants for local planting. These seed potatoes are a very high-value crop with strict control of introducing seed stock certified free of diseases to various countries.

Table 15.1 shows that vine crops of tomatoes, peppers, European cucumbers, and eggplants grow in most hydroponic systems with the exception of water culture and aeroponics. They also do well in most substrates that offer good drainage. While melons are included in the list of crops, they normally are not grown unless they can be trained vertically by support strings. Their production is usually limited to five to six fruits per vine over the cropping period. For this reason, they are not economically feasible to grow commercially except in countries such as Japan, where they demand a very high price in the marketplace (up to $30–40 per fruit).

TABLE 15.1
Crops and Suitable Hydroponic Systems and Substrates

Plant	Hydroponic system									Substrates						
	Water			Ebb and Flow		Drip Irrigation			Aeroponics	Rock	Exp. clay	Perlite	Vermiculite	Rice hulls	Coco coir	Mixes
	NFT	Raft	Wick	Rock	Exp. clay	Rock wool	Coco coir	Bato buck								
Lettuce	x	x	x						x							
Arugula	x	x	x						x							
Basil	x	x	x						x							
Cilantro	x		x									x			x	x
Chervil	x		x									x			x	x
Chives	x		x									x			x	x
Dill	x		x									x			x	x
Marjoram	x		x									x			x	x
Mint	x	x	x						x			x			x	x
Oregano	x		x									x			x	x
Parsley	x		x									x			x	x
Rosemary										x	x	x	x			
Sage	x									x	x	x	x			
Tarragon										x	x	x	x			x
Thyme	x									x	x	x	x			x
Watercress	x	x	x						x							
Bok Choy	x	x	x						x	x	x	x	x	x	x	x
Chard	x	x	x						x	x	x	x	x	x	x	x
Spinach	x	x	x						x	x	x	x	x	x	x	x
Tomatoes				x	x	x	x	x	x	x	x	x	x	x	x	x
Cucumber				x	x	x	x	x		x	x	x	x	x	x	x
Eggplant				x	x	x	x	x		x	x	x	x	x	x	x
Pepper				x	x	x	x	x		x	x	x	x	x	x	x
Melons				x	x	x	x	x		x	x	x	x	x	x	x
Radish				x	x							x	x	x	x	x
Strawberry	x								x			x	x	x	x	x

FIGURE 15.1   Aeroponic growing of seed potatoes. (Courtesy of the Potato Research Center, Universidad de La Molina, Lima, Peru.)

FIGURE 15.2   Aeroponic seed potatoes. Note the small tubers on the roots. (Courtesy of the Potato Research Center, Universidad de La Molina, Lima, Peru.)

Radish is also a very unique crop that is not grown commercially with hydroponic culture. It is a relatively low-value crop and grows well with soil. Radish will grow in many substrates and in ebb and flow systems, so for home use it would be a feasible crop to grow.

Strawberries grow well in many hydroponic systems including NFT and aeroponics. To increase production, they can be grown in an A-frame system of NFT. They are particularly suitable to growing in plant towers with many types of substrates. The plant towers increase the plant density by about 6 times that expected in beds. They are spaced 3 ft apart within rows and 4 ft between rows. There are a number of commercial operations in Florida and even in Colombia and Peru that grow strawberries in plant towers or vertical sacks. Strawberries require good drainage so the substrate and the hydroponic system must meet this need. Plant towers are easy to construct in your home. The design and construction of plant towers was described in Chapter 13.

Herbs also are very productive in plant towers using most substrates. Most will grow for 4–6 months and some, like rosemary, chives, parsley, mint, marjoram, and oregano, will last up to 10–12 months (Figure 15.3). Again, production is increased

FIGURE 15.3 Various herbs in plant towers. (Courtesy of CuisinArt Golf Resort and Spa, Anguilla.)

6–8 times that of normal bed systems. The key to good production and long-lasting crops in plant towers is good light and adequate drainage. The leachate from the towers can be recycled or it can be an open system. For your home, it is best to re-circulate the solution for up to a month before changing it.

16 Environmental Control Components for Hydroponic Systems

Environmental conditions must be maintained at optimum levels to achieve success with any indoor growing. These factors have to be monitored and regulated with equipment having sufficient capacity for the growing area (Figure 16.1). The following factors are controlled: temperature, air circulation, carbon dioxide (CO_2), light, water quality and temperature, nutrient solution pH and electrical conductivity (EC), and oxygenation of the nutrient solution when using water culture systems. The following discussion looks at each environmental factor to control and equipment that can fulfill that need.

TEMPERATURE

Most plants grow well under specific temperature ranges. They require minimum and maximum levels. In general, night temperatures should be cooler than day temperatures. This differential may be as high as 10°F (5.5°C) or more. For cool-season crops such as lettuce, night temperatures should be about 55°F (13°C) and from 60° to 65°F (15.5–18°C) during the day. With warm-season crops such as tomatoes, peppers, cucumbers, and eggplants, suitable temperatures are 65°F (18°C) at night and 75°F (24°C) during the day. Herbs withstand a wider range of temperatures.

To control temperatures, a heating and cooling system is required. In most cases, in your home the normal temperatures you maintain in the house are within the desired temperature ranges of plants. If growing in a cool basement, a supplementary heating system may be a requisite. Electrical space heaters will meet the plant temperature demands. Use a 220-volt heater to save on the electrical demand. Baseboard heaters would also be good. These could be installed together with your house heating system. Artificial lights for providing light will also generate heat, so they may provide more than what is necessary to maintain optimum temperatures. In such a case, a cooling system will have to be installed to extract the air from the growing room.

Cooling by exhaust fan(s) is the standard method. The exhaust fans are installed in the wall with an outlet to the outside. Any such fan needs automatic shutters that close when not operating to assist in preventing insects from entering the growing room. The size of the fans is calculated on the volume of air in the growing room and the temperature differential that must be reduced within the room. An air exchange of the total volume should be at least one per minute if a significant number of lights

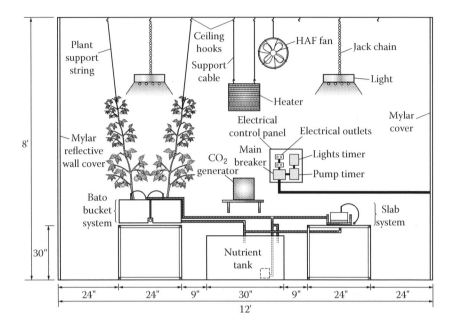

FIGURE 16.1 Cross-sectional view of growing room with all components. (Drawing courtesy of George Barile, Accurate Art, Inc., Holbrook, New York.)

are installed. It is useful to have two-speed fans so that if a small amount of cooling is needed, the lower fan speed operates, but if further cooling is demanded the second faster speed is activated by a thermostat.

It is a good idea to purchase a thermograph that will record temperatures on a 24-hour basis. Some have charts that are good for a week of monitoring before replacing. The thermograph is much better than a max-min thermometer as the thermometer tells you only the maximum and minimum temperatures reached at an undesignated time. It does not tell you temperature fluctuations over time as does the thermograph.

AIR CIRCULATION

Air temperature must be uniform throughout the crop canopy for optimum growth. As the plants grow vertically in the room, they block the circulation of air and cause large temperature differences from the bottom to the top of the plants. This, of course, is especially noticeable with vine crops. The ideal source of heat for the plants is to have floor or bottom heating pipes. Alternatively, a unit heater connected to convection tubes that are located between the rows of vine crops will make hot air rise up through the crop. However, with space heaters a convection tube cannot be used due to the fire hazard. If you mount a gas fired unit heater at one end of the room near the ceiling and install a housing that will distribute the heated air down through convection tubes to the base of the plants, you will get more efficient use of the heat than a space heater. This will also help mix the air and create good circulation making the air temperature throughout the crop more uniform.

The mixing of the air is also important in exchanging the air at the boundary layer immediately in contact with the plant leaves. This movement of air past the leaves brings in fresh air high in CO_2 in contact with the stomates of the leaves where the CO_2 can enter the plant. Another helpful piece of equipment is a horizontal air flow fan (HAF) that is mounted near the ceiling. The HAF fan blows the air down the length of the room causing turbulent flow that mixes the air. This improved air circulation gives more uniform temperature and increases CO_2 exchange at the leaves. Refer to the Appendix for suppliers of this equipment.

CO_2 ENRICHMENT

Normal CO_2 levels in the ambient atmosphere are now approaching 400 mg/L or ppm. Research over the years has established that enrichment of CO_2 to 2–3 times ambient levels (800–1200 ppm) greatly increase the production of vegetable crops by as much as 20%, especially under low light conditions. The ambient CO_2 levels vary with your location. Cities have higher levels of CO_2 due to automobiles and industries. There are a number of small CO_2 generators available for home growers (Figure 16.2). You can use bottled CO_2 in tanks similar to propane tanks, but such a system is awkward and heavy to move for refilling. Also, when using tanks, a system of distribution tubes must be located between and underneath the rows of plants. Another system is a natural gas combustion unit that generates CO_2. This CO_2 generator also gives off heat that may have to be extracted as it operates only during the day period. The operation of the unit is governed by a timer.

The level of CO_2 in the grow room must be monitored and regulated. Many of the generators have a monitor-controller that activates the enrichment system according to preset levels and turns it off if this level is exceeded. You may also purchase a hand-held tester for CO_2 levels.

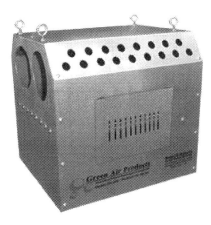

FIGURE 16.2 Natural gas or liquid propane (LP)-fired carbon dioxide generator. (Courtesy of Green Air Products, Inc., Gresham, Oregon.)

FIGURE 16.3   Carbon dioxide Boost Bucket kit compost generator. (Courtesy of CO2Boost LLC, Landenberg, Pennsylvania.)

One of the least expensive and natural methods of generating CO_2 is with a special CO_2 composting pot that releases CO_2 from the decomposition of mushroom compost (Figure 16.3). One bucket will enrich a 10 ft × 10 ft × 10 ft room at levels between 1200 and 1500 ppm for 60–90 days. This is available commercially as a small bucket and is renewable with the purchase of refills (see Appendix for supplier).

LIGHT

Most plants indoors will grow relatively well under a light intensity of 5500 lux (510-foot candles) for a period of 14–16 hours per day. The function of light in photosynthesis and its measurement related to photosynthesis was explained in Chapter 5. In the past, cool-white fluorescent lighting was the type of light used for plant supplementary lighting. Now there are more energy-efficient and better quality lights sold such as the high-intensity discharge (HID) and compact fluorescent. There are two types of HID lights: high-pressure sodium (HPS) and metal halide (MH). A combination of both gives best results. The HPS lights promote blooming and fruiting, whereas, the MH light causes more rapid vegetative growth. Compact fluorescents save electricity. They are now available for growing plants (Figure 16.4). Again, a mixture of the red for flowering and blue for vegetative growing is best.

Today, light emitting diode (LED) lights are gaining popularity due to their energy efficiency. There are a number of types available for growing plants that mix red and blue lights in reflectors to maximize their efficiency in photosynthesis. While these lights last up to 50,000 hours, they are expensive.

All lights are placed in reflective fixtures to maximize the reflection and distribution of the light. There are three shapes of reflectors: parabolic, horizontal, and conical. Parabolic reflectors focus light on the plants by directing the light below

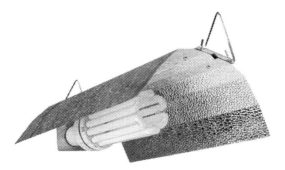

FIGURE 16.4   Compact fluorescent light with light reflector. (Courtesy of Sunlight Supply, Inc., Vancouver, Washington.)

the horizontal plane and thus reducing glare to your eyes. Conical reflectors give more side light. Square-shaped reflectors are effective for square growing areas. Horizontal reflectors are recommended for HPS systems.

Ballasts for HPS lights should be located away from your plants to reduce heat build-up within the crop. Ballasts are not needed for compact fluorescent and LED lights. Keep ballasts off the floor as water could splash onto them. Lights must be mounted above the crop by hooks or jack chains from the ceiling. If you use jack chains, you can raise the lights as the plants grow keeping the lights at least 2–3 ft above the crop. If the lights are too close, the heat they give off will promote rapid vegetative growth.

There are also circular and linear light movers that move the lights above the crop over a period of time to more evenly distribute the light intensity over the crop (Figure 16.5). Linear movers are best for narrow growing areas, whereas circular movers are better for square areas. Light movers do not allow you to plant your crop at higher densities; they improve the light distribution and therefore produce a more even growth of all the plants (Figure 16.6).

Use a Mylar reflective covering on the walls surrounding the hydroponic systems to reflect light back into the sides of the plants.

Refer to the Appendix for suppliers of lights and their accessories.

WATER QUALITY AND TEMPERATURE

Water quality and temperature are important factors for successfully growing plants. Water quality is a measure of the types and concentrations of minerals present in the raw water. Most city waters are of acceptable quality, unless they are very "hard." Hardness is a measure of the amount of calcium and magnesium carbonate present in the water. Hard waters are quite suitable for hydroponics. The only thing you must determine is which elements and at what concentrations they are present so that you may adjust your nutrient solution formulation accordingly. If you purchase a ready-made formulation, ask for one that is for hard water. If you make up your own formulation, hard water may provide most of your calcium and magnesium needs, so will actually save on the use of the fertilizer salts required.

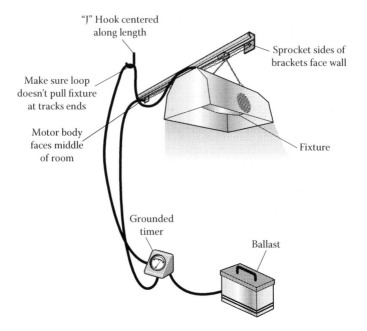

FIGURE 16.5 Linear light mover with components. (Drawing courtesy of George Barile, Accurate Art, Inc., Holbrook, New York.)

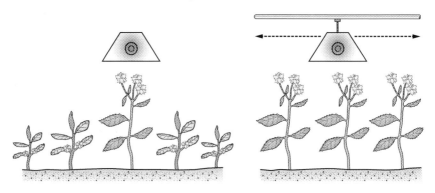

FIGURE 16.6 Comparison of plant growth with and without a moving light source. (Drawing courtesy of George Barile, Accurate Art, Inc., Holbrook, New York.)

The best approach is to get a water analysis of the raw water by a laboratory that does such tests. Test for all of the plant essential elements. Once you know the levels, you can adjust the nutrient formulation to take those into account. When the analysis of the trace elements indicates some are high, reduce or even exclude those from the formulation. For example, optimum boron concentration for the plants is about 0.3 mg/L, so if the test shows boron in excess of that leave out any addition of it in the formulation. Most plants will tolerate boron levels twice or three times the optimum level before any reduction in growth occurs. Often information on water quality may be obtained from the city offices that regulate domestic water supply.

The temperature of the raw water is also a factor that affects plant growth. Everyone is aware that you do not start your soil garden too early in the spring when the soil and groundwater are very cold as the plants will not grow and often just survive without any new growth until the weather heats up the soil. With hydroponics, water temperature is also important for rapid growth. If the water is too cold, purchase an immersion heater. This type of heater is generally electric for small nutrient tanks. They should be available at local hydroponic shops (see Appendix). The immersion heater is placed in the nutrient tank. Temperature is controlled with a thermostat on the immersion heater.

Keep the nutrient solution between 65 and 70°F (18–21°C). Water may also be too high in temperature. In that case, use a water chiller to lower the nutrient solution temperature. For example, lettuce likes a cooler water temperature of 65–68°F (18–20°C). If raw water enters at 80°F (27°C) or more, it is best to lower the temperature to prevent bolting (lettuce going to seed grows a shoot rapidly) and fungal activity in the roots that will damage the plants causing them to wilt during high-light conditions of mid-day. Water holds more oxygen at lower temperature so this also affects plant roots and growth. Alternatively, an ozone generator may be used to add oxygen to water, which will prevent diseases and slow any bolting of lettuce.

Controlling nutrient solution temperature is critical in growing crops in nutrient film technique (NFT) and raft culture systems, especially cool-season crops such as lettuce and spinach. Sources for these components are listed in the Appendix.

EC AND PH OF NUTRIENT SOLUTION

To monitor the concentration of nutrients in the nutrient solution use an EC meter (Figure 16.7). Hand-held "pen" type EC meters are also available (Figure 16.8). These meters tell the total dissolved solutes in the solution and therefore indicate when the

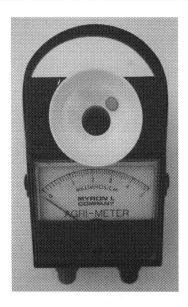

FIGURE 16.7 An electrical conductivity meter.

solution is being depleted of its elements. The addition of elements can be calculated and a "top-up" solution used to increase the elements. However, on a small scale it is much easier to simply replace the nutrient solution making up a new solution every few weeks. Plants use more water than nutrients, so the addition of water is needed every few days (depending upon the ratio of volume of water to each plant). With a

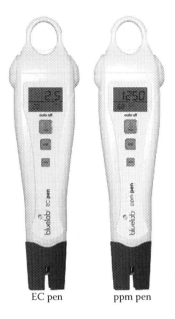

EC pen   ppm pen

FIGURE 16.8   A "pen" type hand-held electrical conductivity and ppm tester. (Courtesy of Bluelab Corporation Ltd., Tauranga, New Zealand.)

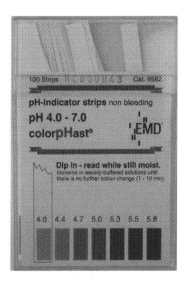

FIGURE 16.9   pH indicator paper.

larger solution tank, the make-up of water and nutrients is less frequent. The addition of water to a nutrient reservoir can be automated using a float valve.

The next factor to monitor is the pH of the solution. Test it every day and add an acid or base to lower or raise it, respectively. You may test it with an indicator paper similar to that used for fish aquariums (Figure 16.9) or a small hand-held pH meter (Figure 16.10). These products are available at hydroponic stores (see Appendix). You can also purchase "pH Up" and "pH Down" solutions from hydroponic outlets or online (Figure 16.11) as mentioned in Chapter 9.

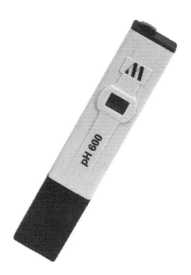

FIGURE 16.10   "Pen" type portable pH meter. (Courtesy of Milwaukee Instruments, Rocky Mount, North Carolina.)

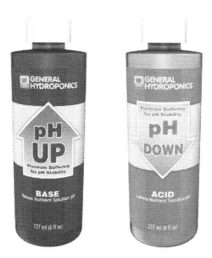

FIGURE 16.11   pH Up and pH Down solutions. (Courtesy of General Hydroponics, Sebastopol, California.)

OXYGENATION OF SOLUTION

This applies mainly to water culture systems. It is, however, helpful for all hydroponic systems to oxygenate the nutrient solution in the solution tank. Achieve this with an air pump and some air stones. These can be acquired from an aquarium shop or online (see Appendix). Place the air stone(s) in the nutrient reservoir connected by poly tubing to the air pump. Keep the air pump elevated above the level of the tank to prevent any possible damage to the pump by water returning to the pump, should the power fail. For raft culture, place a series of air stones along the bottom of the raceway. Connect with poly tubing the air diffusers to an air pump. Many sizes and types of air pumps from linear piston air pumps to diaphragm pumps are available from fish culture suppliers such as Aquatic Eco-Systems, Inc. (www.AquaticEco.com). They also sell tubing, fittings, and diffuser air stones.

Controlling all of these environmental factors at optimum levels will assist in the success of growing hydroponically. Hydroponic culture enables the grower to maximize production and quality. If one or more environmental factors are not at optimum level for the crop, the ability for hydroponics to maximize quality and yield will be restricted.

Section V

*Year-Round Growing
in Greenhouses*

17 The Greenhouse Environment for Plants

LIGHT

A hydroponic backyard greenhouse has more optimum environmental control than in a spare room or basement of your home. One of the greatest limiting factors of indoor home growing is the quality and quantity of light that can be provided by supplementary artificial lighting. The greenhouse uses natural light, and hence benefits greatly from natural sunlight having the quality and quantity of light under which plants evolved. Supplementary artificial lighting is added for winter months to increase light intensity under cloudy conditions and to extend day length during the short days from mid-November through mid-March.

The most common lights to use for this purpose are cool-white fluorescent and high-intensity discharge (HID) lights as mentioned in Chapter 16. The measurement of light intensity in the photosynthetically active radiation was explained in Chapter 5. The amount of light units depends upon the total area of the greenhouse. For example, a 10½ ft × 12 ft greenhouse would need three 8-ft high-output cool-white fluorescent lights, one placed above each bed of plants or six 4-ft T5 fluorescent lights. A 10½ ft wide greenhouse would fit three hydroponic beds. This is a total area of 126 ft^2. Two 600-watt HID lights would provide ample light for this size of greenhouse. They would give about 550 ft candles of light intensity. Mount them a few feet under the roof of the greenhouse above the aisles between the plant beds to get maximum light distribution. They should have horizontal reflectors. Several compound light emitting diode (LED) lights could be used to save electrical consumption, but the initial cost is higher than other light sources. LED lights are available as 4-ft fixtures with either four or eight lamps. The lights can be mixed with red and blue tubes. You would need at least four 4-ft fixtures at a cost of approximately $200 each. Compact fluorescent lights are also available with red and blue mixed lamps in a reflective fixture. The lamps are available in 125-watt, 200-watt, and 300-watt sizes. The lamps and fixtures cost about $100–200 each depending upon their wattage. For a greenhouse of 126 ft^2, at least four 300-watt or six 200-watt units would be needed.

To calculate the amount of lights for a specific area, use a base of 50–70 watts per square foot of growing area. For a 126 ft^2 greenhouse, the actual growing beds would be two beds 2 ft × 12 ft and one bed 9 ft × 2 ft. The total growing area is $2 (2 \times 12) + 1 (2 \times 9) = 48 + 18 = 66$ ft^2. You would need 66×50 watts = 3300 watts. This calculation, however, is for a dark room without natural light; hence, a little less than half of the wattage (1200–1500 watts) would be sufficient for a greenhouse.

TEMPERATURE

Temperature fluctuations in a greenhouse are dependent upon the outside weather conditions, unlike indoor gardening in a house with insulation. The covering of the greenhouse determines how much insulation capacity it has. However, keep in mind that light is extremely important so choose a material that has good light transmission. Common covering materials include glass, polyethylene, and polycarbonate. A double layer of polyethylene will provide more insulation than glass. Polycarbonate is available in various laminates. Double and triple layers are available to reduce heat loss. It also comes in corrugated, but that has less insulation capacity than the double or triple layers. Several disadvantages to polyethylene covering are that even with a special greenhouse film with polymers to reduce ultra violet breakdown, it will last only 3 or 4 years before replacement. Double polyethylene has an advantage of saving energy loss due to a dead air space between the two layers of film, but the disadvantage is the loss of light due to the two layers. This may be as much as 20% or more with aging of the film. Polyethylene covering also does not have the aesthetic appearance of glass or polycarbonate.

Temperature control includes heating and cooling. Due to the "greenhouse effect" as light enters a greenhouse, its wavelength is changed and the heat cannot escape, so there is a heat build-up within the greenhouse under sunlight conditions. When it is cloudy this does not occur. Even during winter months, a very bright sunny day will cause temperatures in the greenhouse to rise above optimum and then some ventilation is needed. During summer months, cooling is a key factor to maintain optimum temperatures during the day when it is sunny. This is achieved by the use of exhaust fans and, if necessary, cooling pads. In backyard hobby greenhouse due to their small area, normally exhaust fans are adequate without cooling pads, unless you are located in a very hot area with desert conditions.

Heating is needed during cool seasons during the days and nights when sunlight is not present. The best form of heat is by use of a boiler and hot water heating pipes at floor level. However, this is not economically feasible with small greenhouses and therefore space heaters are used. The best type of space heater is a unit heater mounted near the ceiling of the greenhouse, but these types of heaters are much more expensive than free-standing space heaters. The most common are electrical and gas- or propane-fired models. If you wish to use an electrical space heater, purchase a 220-volt one that is more energy efficient than a 110-volt one. There are also propane heaters that are located remote from the propane cylinder or others that are mounted on top of the cylinder. More details on heaters, calculations of sizes needed to meet temperature differentials, exhaust fans, covering materials, and so on are given in Chapter 21 on "Components for Backyard Greenhouses."

CARBON DIOXIDE ENRICHMENT

As discussed in Chapter 16, carbon dioxide (CO_2) enrichment increases production for all greenhouse crops by up to about 20%. Small CO_2 generators were discussed in that chapter. I recommend using the same generators as you would for indoor culture for a backyard, hobby greenhouse. However, whenever the greenhouse is ventilating, especially in high-light, late-spring to early-fall periods, a CO_2 generator would not be needed and in fact would waste fuel during that time. Under cloudy weather the CO_2 enrichment is very beneficial, so at that time have the generator operating, but only during the day period, not at night. If you use a propane-fired heater, it will naturally generate CO_2, so the need for an additional CO_2 generator is unnecessary.

RELATIVE HUMIDITY

Optimum relative humidity is about 75% for most plants. High humidity in the greenhouse during cloudy, dull, wet, weather is the main concern. If plants appear moist on their leaves, the best method to reduce the humidity is to heat and exhaust some of the heated air to exchange the air with drier air. Even if it is raining outside and the relative humidity is 100%, normally during such clement weather the ambient temperatures will be low, so heating of the greenhouse air is taking place. As the heater raises the incoming colder air, the humidity will fall. Because of this physical relationship between rising temperatures and falling relative humidity, as the greenhouse air is exhausted and replaced by cold outside air that is heated as it enters the greenhouse, the relative humidity is reduced with the incoming air.

During summer, with hot, dry weather, the greenhouse air may become very low in relative humidity. In this case, you can install some misters above the crop rows. They would be at about 3 ft centers, depending upon the specifications of the mist nozzles, along a ¾″ diameter PVC pipe. They are attached to the pipe by special "saddles" glued to the pipe. The header pipes are connected to a pump capable of 40–60 psi pressure to attain a fine mist. The system is operated by a time-clock or other controller. Misting cycles and periods depend upon the level of relative humidity. The lower or drier the conditions, the more often the mist is activated. Normally, this would be about every 15–20 min for a period of 10–15 sec. If the relative humidity is very low, plants may lose excess water causing the stomates to close, restricting the entrance of CO_2 and therefore slowing photosynthesis and plant growth. Plants under such conditions can wilt and leaf margins may burn.

IRRIGATION

Drip irrigation is with a pump in the nutrient tank that is activated with a time-clock or controller. The cycles (time between irrigations and the length of time during an irrigation cycle) depend upon the weather conditions (temperature, sunlight, relative humidity), season (day length), crop stage of growth, and the nature of the crop. Low-profile plants are normally grown in a nutrient film technique (NFT) or raft culture system so irrigation is constant. Vine crops growing in a substrate with drip irrigation require closer monitoring of irrigation cycles and adjustments according to the earlier-mentioned factors. Large-leaf plants such as European cucumbers and eggplants due to the expansive leaf area surface require more irrigation than do tomatoes or peppers having smaller total leaf area. All of these parameters of caring for your plants will be a learning experience, especially for the first crops undertaken.

However, that is the fascinating challenge of growing that will increase yields with your experience of tending to them.

Time-clocks and controllers suited for drip irrigation are discussed in Chapter 21. Whenever an irrigation cycle occurs for hydroponic systems, irrigation is always with a nutrient solution, not raw water. If plants wilt during mid-day, high-light conditions, the probable cause is lack of water. Immediately increase the irrigation cycle frequency. The duration of an irrigation cycle is also important as mentioned earlier in Chapter 6 in achieving sufficient leachate to prevent any salt build-up in the substrate.

18 Benefits of a Greenhouse to the Homeowner

I personally became involved in backyard greenhouses in the mid-1970s. I and an engineer friend started a company, Resh Greenhouses Ltd., in Vancouver, Canada. We had been working part-time on commercial hydroponic greenhouses near Vancouver and thought that building small hobby greenhouses equipped with full hydroponic systems may have a market for homeowners. At that time I was a full-time student doing research for my doctorate degree and my engineer friend had his civil engineering consulting business. We became very enthusiastic about introducing hydroponics into society at the level of backyard gardeners. The science was fairly new at that time in the commercial greenhouse industry growing vegetables. It was our opinion that most people would like to grow vegetables in a cleaner environment than soil and at the same time reduce the stooping labor of growing in their gardens. Added to that, growing in a greenhouse would extend the growing season to year-round.

Our company manufactured small hobby backyard greenhouses that were prefabricated from aluminum framework and fiberglass covering. The package came with all components of heating, cooling, and a hydroponic system. At that time gravel culture was the easiest method of hydroponics. Located in a temperate climate that had short, rainy, cold weather from late fall to early spring, backyard greenhouses would offer a new dimension to gardening during those seasons when no traditional outside gardening was possible. Most people really enjoyed the presence of houseplants in their homes year-round, but were not able to grow vegetables inside due to the messiness of soil growing. People working all day at their normal jobs would leave the house in the early morning when it was still dark and return in the evening under dark winter conditions. They missed the outside activity of gardening during the winter months. Our backyard greenhouses were equipped with artificial supplementary lights that extended the day length to 14 hours, so when clients returned from work in the evening they would go into the greenhouse and live summer conditions with all of their plants growing. Most of our customers integrated the rejuvenation of house plants into the greenhouse with the vegetable crops and soon wished that they had a larger greenhouse.

Hydroponic culture had a fascination in itself for the clients. Clients wished to learn more about the nutrients, different hydroponic growing systems, plant care, and pest management. The interest grew and soon I became involved in teaching extension classes on hydroponics during the evenings at the University of British Columbia. Many of the students were owners of our greenhouses. The students often asked me to write a book on hydroponics as there was no real compiled information from one source, but just bits and pieces from various papers and journals. That was the birth of my first book, *Hydroponic Food Production*, in 1978.

Growing hydroponically offers very clean conditions without any soil. Plants inside the backyard greenhouse are protected from many insects and diseases and keep the vegetables away from contact with soil. The vegetables are grown to full mature stage, so flavor and nutrition is beyond what could be experienced in store-bought products. This heightened flavor and nutrition of healthy products grown by the hobbyist is very self-satisfying. People with backyard greenhouses produce lettuce, herbs, tomatoes, peppers, and cucumbers out of season to backyard outside gardening. Lettuce and herbs can be grown constantly by sowing and transplanting new crops weekly, whereas the vine crops are seeded in October as the outside garden comes to an end. With artificial lights in the greenhouse (cool-white or T5, high-output, fluorescent) located above the beds (3) of plants, it is possible to get tomatoes in production by January and peppers by February. Production continues until mid-summer when outside gardening provides these same crops. The greenhouse crop is changed during August–September, the greenhouse cleaned, and a new crop started by October again.

If the hobby grower wishes to grow other vegetables such as radish, chard, arugula, green onions, bush beans, and others, he can easily do so in the greenhouse. The type of hydroponic system to use then depends upon the crop as we discussed in Chapter 15. Lettuce, arugula, basil, and some other herbs do best in nutrient film technique (NFT), while root crops such as radish would be in a peatlite mix. Having these salads available fresh during winter months is a unique experience for backyard growers. They may share them with family and friends imparting a feeling of pride and accomplishment that others cannot do without a backyard greenhouse (Figure 18.1).

FIGURE 18.1   Happy gardener harvesting vegetables from his hydroponic backyard greenhouse. (Drawing courtesy of George Barile, Accurate Art, Inc., Holbrook, New York.)

The transition from a wintery environment outside to a tropical paradise inside the greenhouse with light and pleasant temperatures offers an escape from the depressing ambient weather conditions. It lifts the hobbyists from their reality of work and darkness into an environment not experienced at that time of the year. This psychological effect on people with the surroundings of plants raises their spirits to be more positive and energetic with their normal daily routines. The greenhouse is a sanctuary, a special garden retreat from the everyday hustle and bustle. With your senses bathed in solitude with light and abounding aromas of live plants, you relax as you tend the plants or simply sit and savor the sweet smells of the surrounding nature. It is a place to unwind, read a book, or enjoy a cup of tea conversing with a friend or your soul mate. The greenhouse environment takes you on a mini-vacation to a tropical paradise.

Hobby hydroponic greenhouse growing is very educational for children, introducing them to the miracles of nature in the growing of food. They learn that their food does not grow on supermarket shelves. They will be thrilled by the development of plants from seed to producing vegetables in the environment of the greenhouse. Students can carry out science projects on plant nutrition and nutrient solutions applying their school chemistry to something that actually materializes in food production. Overall, it will give them an appreciation for the amount of work involved in producing crops and teach them some self-reliance of growing their own vegetables in the greenhouse, and later during spring and summer they can extend their interest to having their own vegetable garden outside. They can start their plants in the greenhouse as bedding plants and later transplant them to their outdoor vegetable garden. Of course, this also applies to the hobby greenhouse grower who can start all of his or her flowers and vegetables from seeds in trays in the greenhouse and plant them outside when temperatures are favorable. Sow tomatoes, peppers, eggplants, herbs, and flowers of your varietal choice in the greenhouse during March to April to begin your outdoor gardening in May.

Owning a backyard greenhouse is an investment in your enjoyment of nature, your peace of mind, and your family's health!

19 Design, Layout, and Construction of Backyard Greenhouses

When selecting a backyard greenhouse consider first the size. The size of greenhouse best suited for your needs depends upon what you wish to grow and your weekly consumption. From past experience with manufacturing greenhouses, I found that a 10½ ft by 12 ft greenhouse could easily supply a family of four to six persons. Most clients soon started using their greenhouse for rejuvenation of their indoor house plants so took some of the greenhouse space for them. Our 8 ft by 12 ft greenhouse had about one-third less growing space since only two beds would fit instead of three with the 10½ ft wide model. In terms of cost and growing space, we found that the 10½ ft by 12 ft model was the best value for the hobbyist.

GREENHOUSE LOCATION, TYPE, AND SHAPE

Hobby greenhouses may be free-standing or lean-to structures (Figure 19.1). The choice here depends upon the location and space available in your backyard. A lean-to structure can be attached to a building wall such as your garage or onto a tall wooden fence. Attaching it to a building has the advantage of getting some insulation from the building. Keep in mind when locating such a lean-to structure that it must receive adequate natural light so the location must not be in a shaded area or on the north side of a building that would prevent natural sunlight from entering. On a fence do not locate it on a south-side fence where the fence would restrict direct light to the greenhouse.

The best location in your backyard is where the greenhouse can receive sunlight most of the day (Figure 19.2). To achieve this locate it in the middle or at one side where no shade is cast by fences, buildings, or trees. If your backyard is fully treed or large trees in neighboring yards shade the light entering your yard, the trees will have to be thinned out by removing lower branches or take out the trees. If the shade is cast by neighboring trees you will need cooperation from the adjacent property owners to prune their trees or perhaps remove some of them. You may be able to convince them to do so if you are willing to share some of your vegetables with them.

A free-standing type of greenhouse is most efficient in receiving sunlight from the east, south, and west cardinal directions. It also gives you the greatest growing area for its size. If possible, orient the greenhouse north-south. That is, the ridge of the structure is oriented north-south. This will give the best light for vine crops as light will enter the rows of plants also oriented north-south. If oriented east-west, there will be mutual shading of plant rows with the south facing row receiving most light as it will cast

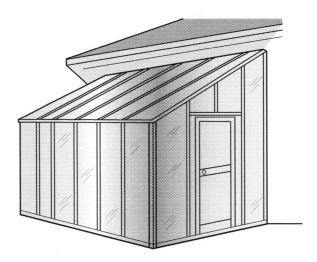

FIGURE 19.1 Lean-to greenhouse attached to a building. (Drawing courtesy of George Barile, Accurate Art, Inc., Holbrook, New York.)

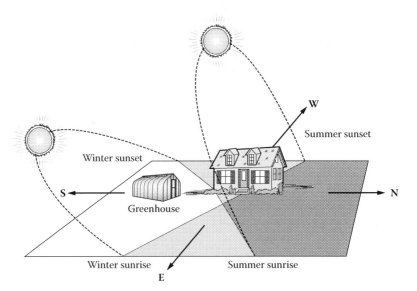

Locate the greenhouse where there is a minimum of shade cast by trees, fences and buildings.

Poor location: some shade almost all day

Good location: some shade either in the morning or the afternoon

Best location: no shade

FIGURE 19.2 Location of backyard greenhouse according to sun's movement. (Drawing courtesy of George Barile, Accurate Art, Inc., Holbrook, New York.)

FIGURE 19.3   Traditional gable, free-standing greenhouse. (Drawing courtesy of George Barile, Accurate Art, Inc., Holbrook, New York.)

shadows on the other rows, especially during the fall, winter, and early spring months when the sun's angle of incidence is much lower than during the summer months. Pick the spot in your backyard that has best sunlight exposure to maximize crop production.

Whether the greenhouse is a lean-to or a free-standing structure, there are two basic shapes. The traditional gable with straight eaves where the sides meet the roof and the gothic shape with curved eaves where the sides bend to form the roof (Figures 19.3 and 19.4). While you may construct a backyard greenhouse with a polyethylene covering, it lacks the beautifying appearance for your backyard compared to the clarity of glass or polycarbonate. Greenhouses with curved eaves use polycarbonate to enable forming the bend at the eave location. Glass is perfect for the traditional gable-shaped greenhouses, but polycarbonate is also suitable. Polycarbonate has the advantage of higher insulation. If you wish to grow year-round in your greenhouse it must withstand wind, snow loads, and resist ultraviolet (UV) breakdown of the covering. Polyethylene has a limited life expectancy due to UV light turning it brittle as the polymers are damaged. The last thing you want in your backyard is an eye-sore structure that turns milky to a yellowish color with age of the covering. Also, if the covering has to be replaced in the cold winter months you risk losing all of your plants. For these reasons, invest a little more in a permanent structure that has a nice appearance and will withstand the weather all year.

NATURE OF GREENHOUSE STRUCTURE

Some do-it-yourself (DIY) papers suggest that you construct a greenhouse using plastic polyvinyl chloride (PVC) pipe. It combined with wooden end-framing allows you to construct a Quonset style house, but this type of structure will not withstand snow loads or strong winds. These types of houses are okay as cold frames using polyethylene covering. They are really only cold frames suited to starting your bedding plants early in the spring. They are not greenhouses with heating and cooling systems that permit you to grow year-round.

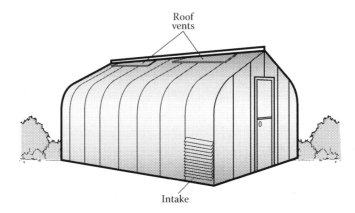

FIGURE 19.4 Free-standing gothic greenhouse. (Drawing courtesy of George Barile, Accurate Art, Inc., Holbrook, New York.)

You may construct the greenhouse of treated wood framing, but that is not as permanent as aluminum framework. With wood framing you could cover it with polycarbonate, but not easily with glass as that would require a lot of framing for the glass panels. In the end, you may not save much money building it yourself. I recommend that you purchase a hobby greenhouse in the market as there are many sizes and shapes available to fit your budget and also most are constructed of aluminum framework so are permanent for many years. In addition, the aluminum frame comes in white, green, or brown so you can pick a color that adds to the overall appearance of your home and backyard. Galvanized steel structures are fine, but may develop rust spots over time. I do not think that they are much cheaper than aluminum structures. The aluminum will retain its beauty over time so will always be an asset to the surroundings of your home.

The frame of the greenhouse must be strong enough to support your plants whether they are hanging baskets or vine crops supported by string to an overhead cable attached to the greenhouse superstructure. It must also support the covering and equipment such as heaters and fans. If it is wood, the structural members or rafters, as in your house, must be sufficiently close together to prevent any sagging of the covering material, especially in the winter under snow loads. You may need to attach cross members to the rafters in several locations to strengthen the support of the covering material, such as glass or polycarbonate. Additionally, wooden structures intercept light causing a lot of shading, especially in the winter when you need all the sunlight available. If you use wood, be sure to paint it with white paint, preferably an oil-based paint to resist the constant moisture present in the greenhouse. Aluminum framing naturally reflects light and at the same time the dimensions of the ribs and purlins (cross members) are much smaller so do not cast as much shadow as wood framing.

SIZES AND PRICES

There are many sizes of both lean-to and free-standing greenhouses. The most common dimensions for lean-to greenhouses are in widths of 6 ft, 8 ft, and 10 ft with lengths of 6 ft, 8 ft, 10 ft, and 12 ft, with increments of 2 ft to 20 ft or longer. For the best utilization of the growing space choose a 10 ft width with 12 ft in length or more. Prices for these houses range from $3000 to $6000. Prices also depend upon the covering material. For example, double tempered glass is the most expensive, elevating prices to $15,000 for the larger structures. Most come with two roof vents operated by solar vent openers that do not require electricity. These vents open as sunlight builds up heat around a cylinder mechanism that pushes open the vent.

Free-standing greenhouses come in widths of 8 ft, 10 ft, and 12 ft and in 2-ft increments to 20 ft. Similarly, lengths are available in 2-ft increments starting at 10 ft up to 20 ft and longer. I feel the most common size for a family of four to six members is 10 ft by 12 ft or 10 ft by 16 ft. That will produce sufficient vegetables to please the entire family. Prices for these greenhouses range from $4000 for an 8 ft by 12 ft to $6000 for a 10 ft by 12 ft and higher. The covering material also influences prices as mentioned earlier.

COVERING (CLADDING)

The more light let through (transmissibility) the covering, the better for plant growth. Coverings include glass (single or double tempered); rigid plastic (polycarbonate)- single, double wall, triple wall, five wall; and polyethylene-single or double layer with dead air space between the layers. The covering type influences not only the price, but most importantly the insulation capability.

Glass is the favorite covering for gable style greenhouses. It is very attractive, showing plants growing visible from the outside and offers lots of light, especially during the winter. Single and double tempered, unbreakable glass is available. The double glass retains heat better than the single layer. If you purchase a single layer glass greenhouse you can insulate it with a layer of polyethylene on the inside during the winter to retain heat. This, of course, will reduce some of the incoming light and make the greenhouse less transparent so it may lack the visual appeal of the glass alone. I do not like this method to retain heat as the inside poly layer will also cause more condensation that can drip on your plants and promote potential diseases. It is far superior to purchase the tempered double glass covering. One thing that can be done to reduce some heat loss is to install 2″ thick Styrofoam on the greenhouse sides below the bench level. In most cases the hydroponic systems for lettuce nutrient film technique (NFT) and vine crops use raised trays so that they are at least 2 ft above the floor. Thus, the Styrofoam will not significantly reduce light at the crop level if it is fitted to 2 ft high around the inside of the greenhouse walls. Regardless of the covering, the use of the Styrofoam along the base walls will save additional energy loss.

Polycarbonate is my choice of covering as it is available in different wall thickness to reduce heat loss. Polycarbonate is available in 4 mm, 6 mm, and 10 mm twin wall and 8 mm and 16 mm triple wall thicknesses (Figure 19.5). A 16 mm 5-wall poly-carbonate gives the best insulation, but its light transmission is reduced to 62%. This would not be good for regions having very dark winters with little natural sunlight. For example, in the rainy areas of the West Coast where minimum winter temperatures are not that low, use the twin-wall polycarbonate. In Central regions and the East Coast where extremely cold periods occur during the winter, but in most cases under sunny conditions, the thicker polycarbonates of triple wall and five walls would be okay.

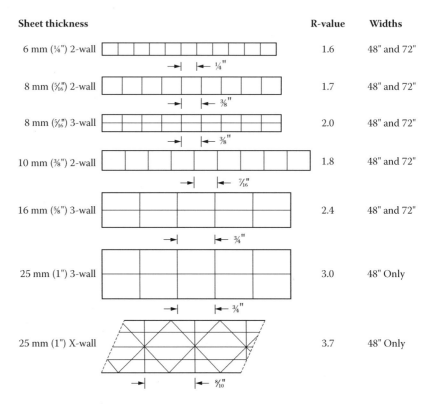

Sheet thickness		R-value	Widths
6 mm (¼") 2-wall	¼"	1.6	48" and 72"
8 mm (⁵⁄₁₆") 2-wall	⅜"	1.7	48" and 72"
8 mm (⁵⁄₁₆") 3-wall	⅜"	2.0	48" and 72"
10 mm (⅜") 2-wall	⁷⁄₁₆"	1.8	48" and 72"
16 mm (⅝") 3-wall	¾"	2.4	48" and 72"
25 mm (1") 3-wall	¾"	3.0	48" Only
25 mm (1") X-wall	⁸⁄₁₀"	3.7	48" Only

FIGURE 19.5 Comparison of multi-walled polycarbonates. (Drawing courtesy of George Barile, Accurate Art, Inc., Holbrook, New York.)

Light transmission of polycarbonate is diffused and twin-wall polycarbonate light transmission is 10% less than that of glass. The effectiveness of thermal insulation of a covering is measured as an R-value the same as insulation in your home. The higher the R value the greater the heating and cooling efficiency. Here are a few R-values for different greenhouse coverings:

Product	R-values
Single 6-mil polyethylene	0.87
Double 6-mil polyethylene	1.70
3 mm single tempered glass	0.95
Double tempered glass	2.00
4 mm twin-wall polycarbonate	1.43
6 mm twin-wall polycarbonate	1.54
8 mm twin-wall polycarbonate	1.72
10 mm twin-wall polycarbonate	1.89
16 mm triple-wall polycarbonate	2.50
16 mm 5-wall polycarbonate	3.03

Another aesthetic plus for polycarbonate is its ability to be bent to form the curved eaves making the gothic shape of a greenhouse. This has a very attractive modern look.

The greenhouse door should be of aluminum and tempered glass window with a screened opening that allows ventilation during the hot weather. The sidewall height should be at least 6 ft to allow the training of vine crops up to the roof height. Additional height is another advantage of the curved eaves of the gothic shape. The ridge height should be about 8–9 ft. To raise the greenhouse height you may place it on top of a brick or masonry concrete block foundation. This can raise the greenhouse up from 16″ to 24″. Just realize that the door must then be positioned this same height lower than the greenhouse base so that there is not a step into the greenhouse.

SITE PREPARATION

The greenhouse should be easily accessible and be located within reasonable distance from a source of water and electricity. A PVC water line will have to be buried below the expected frost depth of winter months, usually at least 3 ft. A 1″ diameter main will be adequate. When digging the trench add an underground buried cable that will carry a 220 volt, 80–100-amp service from your house to the greenhouse. These will be stubbed at a location on the back wall of the greenhouse, so a good plan should be made to get the exact location of the greenhouse. Once the utilities are in place, lay a landscape weed mat over the entire area that the greenhouse will occupy and extend it several feet beyond the perimeter of the greenhouse. This is to prevent weed growth within and immediately adjacent to the greenhouse.

In the center of the weed mat construct a treated 2″ × 4″ wooden frame having the edges of the wood laying on top of the weed mat (Figure 19.6). For example, for a 10 ft × 12 ft greenhouse make a wooden frame 12 ft × 14 ft. Attach the corners of the frame with metal brackets screwing them into the wood with stainless steel screws. Level and secure the frame in place with 1″ × 2″ × 18″ treated wood stakes or rebar at the corners and along the sides. Be sure that the frame is square by measuring the diagonals to make them equal length. Place 1″ thick Styrofoam on the bottom for insulation. Then fill the entire framed area with pea gravel to give the greenhouse a good base with free drainage. The greenhouse will then be set on top of the pea gravel. If you wish to get a very nice clean look, install stepping stones in the greenhouse and as a path to the greenhouse from your home.

Some references recommend that the top ridge of the greenhouse be oriented east to west. This will provide maximum sunlight exposure during the winter when the angle of incidence of the sun is very low; however, this applies only to low-profile crops and potted ornamental house plants. For vine-crops, orient the greenhouse ridge north-south so that the sun will shine down the rows of plants and not just maximize light on the south row as it will shade the others.

When erecting the greenhouse, fasten its base to 6″ × 6″ treated wooden timbers or make a foundation with several layers of 8″ × 8″ × 16″ concrete blocks. The blocks must sit directly on the weed mat, not on top of the pea gravel fill. The use of blocks will raise the height of the greenhouse. If you use treated wooden timbers for the base install a 10 mil polyethylene barrier between the base and the aluminum greenhouse sill to prevent corrosion of the aluminum.

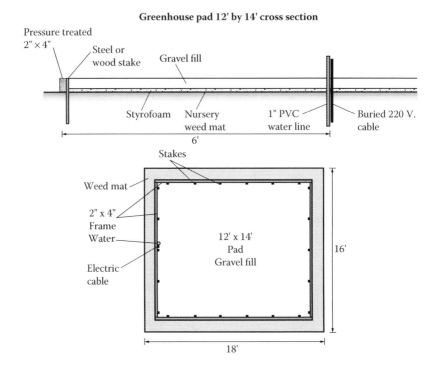

FIGURE 19.6 Site preparation for a backyard greenhouse. (Drawing courtesy of George Barile, Accurate Art, Inc., Holbrook, New York.)

Do not use old railway ties for the base as they are treated with creosote that gives off fumes, especially under warm temperatures, that are toxic to plants. Pentachlorophenol was an early wood preservative, but is now restricted in its use and is not approved for residential application. It gives off damaging fumes so do not use it for greenhouse wood preservation as it acts as an herbicide. Cuprinol, copper naphthenate, is probably the most commonly used wood preservatives used by gardeners. It is recommended for use on sheds, fences, garden buildings, decking, and garden furniture as a wood preservative. It is approved and safe to use on these structures as well as for greenhouses. I would, however, recommend painting the treated wood to seal it and get good light reflection.

CONSTRUCTION OF A GREENHOUSE

The greenhouse frame consists of ribs set onto a sill and purlins and girts (horizontal bars) that bolt to the ribs (Figure 19.7). If this is a prefabricated kit, the sills will screw or bolt into the base you prepared of timber or concrete blocks. Start with the end sides (traditional) or ribs (gothic) and bolt them with the purlins at the eave height (traditional) or ridge in the case of the gothic style. Then attach the rest of the cross members (purlins) to the gable ends. Square the structure and tighten the bolts of this outside frame before filling in the rest of the studs (traditional) or ribs (gothic).

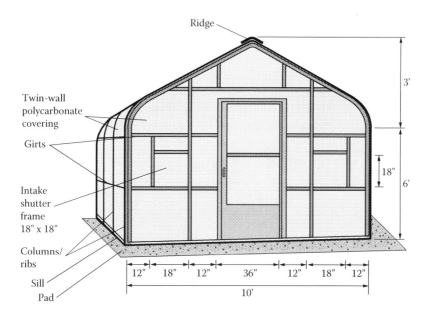

FIGURE 19.7   Greenhouse frame c/w ribs, purlins, gables, vent, and fan openings. (Drawing courtesy of George Barile, Accurate Art, Inc., Holbrook, New York.)

Most structures will have knee braces that keep the house from folding in accordion motion when strong winds push on the gable ends. They attach to the gable end ribs and go on an angle to the sill. They will also keep the house square while attaching the remaining ribs and purlins. With the traditional form, the rafters are attached to the studs at the eave height and are secured at the top with the ridge purlin. With the gothic style, the ribs are both the studs and rafters in one piece and so are attached with the ridge purlin and others (usually two to three) in the roof area. Once you complete the basic frame with the entire purlins, frame the gable ends with the studs. If you plan on using an exhaust fan ventilation system, frame the area where the fan is to be installed on the gable opposite the door, and on the other gable, where the door is located, frame two openings for the automatic shutters. The exhaust-fan frame should be about 5–6 ft above the floor (at the eave level) and the automatic shutters at about 4 ft high. Some backyard greenhouse builders may pre-assemble the gable ends. That will save lots of time in the erection of the house. If the greenhouse has roof vents assemble the framework for them.

The next step is to install the polycarbonate or glass covering. The following description is for the installation of polycarbonate as glass is more specialized and requires that you follow the installation instructions exactly as provided by the greenhouse manufacturer. In any case, if you purchase a pre-fabricated greenhouse, it will come with full step-by-step instructions for its assembly. Complete all the supporting structure (superstructure) and painting before installing the covering. Orient the sheets with the ribs (cells) parallel to the rain flow (slope) and the outside face is the side of the panels marked "UV side out" (Figure 19.8). This orientation applies to multi-walled and corrugated polycarbonate. Polycarbonate can be cut with a jigsaw

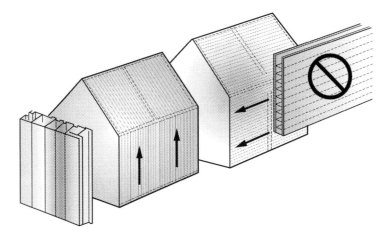

FIGURE 19.8 Installation of polycarbonate sheets. (Drawing courtesy of George Barile, Accurate Art, Inc., Holbrook, New York.)

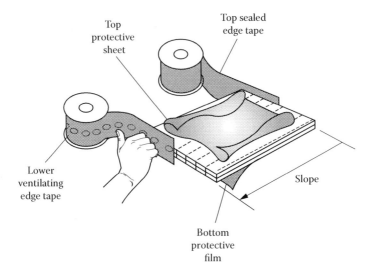

FIGURE 19.9 Tape ends of polycarbonate sheets. (Drawing courtesy of George Barile, Accurate Art, Inc., Holbrook, New York.)

with a fine-toothed blade. Remove any dust in the flutes (channels) with an air compressor or vacuum cleaner. Remove the protective film a few inches back from the edges where you are working with its installation. After installation is complete remove all the protective film. Seal the upper edge of the sheets with a special aluminum tape and the bottom edge with a polycarbonate vent tape that allows moisture to exit but prevents debris from entering into the flutes (Figure 19.9).

Polycarbonate expands and contracts with temperature, so when fastening it to the framework use slightly larger holes than the diameter of the fastener to allow some movement. Set the first sheet on square and screw every fourth corrugation to each purlin of the greenhouse structure. Before attaching the next sheet run a bead of caulking compound, such as silicone, along the last corrugation where one corrugation of lap takes place with the next sheet to seal the laps. At the sill insert a closure strip (foam having configuration of the corrugation) as you screw through the corrugated sheet into the sill. Likewise, at the roof peak insert a closure strip as the sheets are held in place with the ridge bar (Figure 19.10). If the greenhouse is of a traditional shape, closure strips will need to be placed at each end of the panels. At the sill, purlin where the sidewall meets the roof (eave), the lower end of the panel of the roof where it overlaps the sidewall and at the peak use closure strips to seal the panel ends. Allow at least 2″ of overlap of the roof panels beyond the sidewall panels at the eave to make water flow past the sides of the greenhouse. All fasteners should be stainless steel screws or teck (self-drilling) screws (Figure 19.11) that have neoprene washers under the heads to seal the screw holes as they are tightened to the polycarbonate. Be careful not to tighten the screws excessively causing the polycarbonate to become indented underneath.

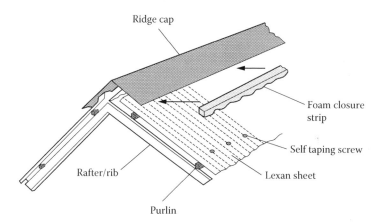

FIGURE 19.10 Installation of closure strips at roof peak of greenhouse. (Drawing courtesy of George Barile, Accurate Art, Inc., Holbrook, New York.)

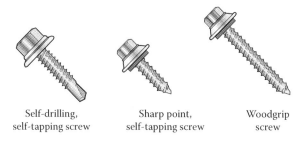

Self-drilling, self-tapping screw

Sharp point, self-tapping screw

Woodgrip screw

FIGURE 19.11 Use of self-drilling teck screw and wood screw fasteners. (Drawing courtesy of George Barile, Accurate Art, Inc., Holbrook, New York.)

If the greenhouse is a beamed structure having 2″ × 4″ rafters without purlins, such as in the case of a wooden structure, spacer blockings will have to be fastened between the rafters (Figure 19.12). The spacing of the rafters and spacer blocks vary with the polycarbonate sheet thickness and the snow load, for example, using 8 mm thick polycarbonate panels, space rafters every 2 ft with blocking spacers at every 6 ft going across the rafters for a roof load of 60 lb per sq ft. With 16 mm thick polycarbonate sheets, rafters may be spaced at 4 ft with blocking every 2.5 ft for a 60 lb load. If the rafters are at 2-ft centers with 16 mm thick panels no blocking is needed. The panel width is 4 ft. Keep the spacer blocks 3/8″ below the rafter face to permit condensate movement past the blocks.

If the greenhouse superstructure is made without purlins, use specific fastener bars to join the polycarbonate sheets (Figure 19.13). These aluminum base fastener bars are attached with self-drilling screws (teck screws) about every foot to the center of the rafters. Work one sheet at a time going across the structure. Slide the sheet under the first fastener bar and then go to the next rafter and slide the fastener base under the sheet and attach it to the rafter. Repeat this with all sheets until complete; then attach the caps to the base fastener bars using a rubber mallet. The caps lock the panels into the fastener bars. Alternatively, you may use just a cap and screw it directly into the rafters without a base bar. Attach the panels to the spacer blocks between the rafters with a screw, washer, and ⅜″ thick by 1″ diameter neoprene spacer. This permits the condensate moisture to flow on the inside of the polycarbonate sheet without it touching the cross blocking so drip will not occur within the greenhouse. To seal the ridge use a ridge cap formed from aluminum. Seal the panels

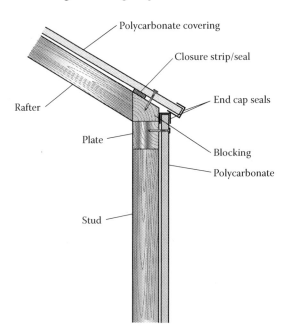

FIGURE 19.12   Spacer blocks with rafters of a wooden greenhouse. (Drawing courtesy of George Barile, Accurate Art, Inc., Holbrook, New York.)

Cap

Base fastener bar

(a) **Attach base bar to rafter/roof beam**

(b) **Place covering sheet on top of base bar**

Beam/
column

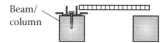

(c) **Slide next base bar under
sheet and attach to beam**

(d) **Attach cap to base bar by
hammering with a rubber mallet**

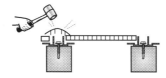

(e) **Add sheets sequentially using Steps (a) to (d),
then, remove protective film after completion**

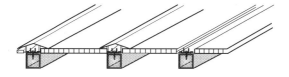

FIGURE 19.13 Installation of polycarbonate to structures without purlins. (Drawing courtesy of George Barile, Accurate Art, Inc., Holbrook, New York.)

into the ridge cap using a foam gasket similar to the closure strips for the corrugated material, but this weather seal is square without corrugations.

If the greenhouse is of aluminum or other metal framework, purlins will be bolted across the ribs. The polycarbonate sheets are attached to the purlins using base and cap bars (Figure 19.14). With self-drilling stainless steel screws fasten the

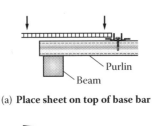

(a) **Place sheet on top of base bar**

(b)   **Attach base bar to purlin**

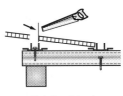

(c)   **Cut sheet to size**

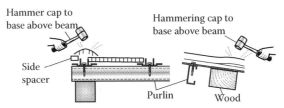

(d) **Side spacer used when hammering cap to base**

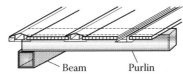

(e)   **Continue adding sheets, etc.**

FIGURE 19.14   Installation of polycarbonate to structures with purlins. (Drawing courtesy of George Barile, Accurate Art, Inc., Holbrook, New York.)

base connecting bar to the purlins underneath the sheet edge. Place these bars at spacing equal to the width of the polycarbonate sheets. Attach the base connecting bars working across the greenhouse as the polycarbonate panels are added to get the correct spacing of the bars. Continue adding all of the sheets and the caps for the connecting bars as you proceed. Underneath the base bar use a shock-absorbing support such as a wooden block as the cap is connected to the base using a rubber mallet when the panel is in position. Remove the protective film from the polycarbonate

TABLE 19.1
Bending Radii of Polycarbonate Sheets

Sheet thickness (mm) (in.)	6 (¼″)	8 (⁵⁄₁₆″)	10 (⅜″)	16 (⅝″)
Smallest bending radius (m)	1.05	1.40	1.75	2.80
Smallest bending radius (ft)	3.44	4.6	5.73	9.2
	3 ft 5 in	4 ft 7 in	5 ft 9 in	9 ft 2 in

and drill screws with metal and neoprene washers to further fasten the panels to the eave purlin at the bottom edge of the panels if it is a gable shape greenhouse. Use a no-drip spacer neoprene washer between the sheet and purlin to allow condensate drainage. The connecting base bars have slightly raised edges to keep the panels from touching the purlins and thus permitting movement of condensation moisture. Once the gable ends, sides, and roof have been constructed caulk the edges of the gable ends where they meet the sidewall sheets.

The various fastening bars, caps, and ridge cap will be supplied by the greenhouse manufacturer; otherwise, if doing it yourself these bars should be available where you purchase the polycarbonate panels. Similarly, foam closure strips, all hardware come with the greenhouse kit or purchase it with the polycarbonate sheets.

When constructing a gothic style house, where the polycarbonate is bent from the walls to the roof, keep in mind that the polycarbonate sheets have a minimum radius that they can be bent without buckling. Table 19.1 shows the smallest cold bending radius for various panel thicknesses.

EXAMPLE OF GREENHOUSE CONSTRUCTION

Here I am presenting an example of the construction of an 8 ft × 12 ft backyard greenhouse. These greenhouses were constructed of aluminum framing and corrugated fiberglass panels as at the time polycarbonate panels were not common and very expensive. That was in the mid-1980s. The procedures, nonetheless, are the same as when they were constructed of corrugated polycarbonate. For twin-wall or multi-wall polycarbonate, the general procedure is the same with some modifications as to the use of the aluminum base bars and caps that attach to the purlins, which was explained earlier. The use of these fastener bars makes the installation simpler than previously as we did in the past with Resh Greenhouses Ltd. using pop-rivets and tech screws onto the purlins directly.

We built our greenhouses with prefabricated finished gable ends to make the erection simpler. However, if the gable ends are not completely finished with cladding, a door, and so on, this can be done after the framework is assembled as was explained earlier. Today, use the twin- or multi-walled polycarbonate panels to cover the entire greenhouse—sides, walls, roof, and end-wall gables. With the base pad ready prepared (Figure 19.6) as discussed earlier, position the treated wooden sills, level and square them, then fasten them with aluminum brackets at the corners. Assemble the ribs, purlins, ridge cap, and gable-end framing, square the entire structure, and tighten all bolts (Figure 19.15). Next, place all of the polycarbonate

FIGURE 19.15   Assembly of gable ends, with door and exhaust shutter, to purlins. (Courtesy of Resh Greenhouses Ltd., Vancouver, British Columbia, Canada.)

FIGURE 19.16   Attaching covering to greenhouse structure. (Courtesy of Resh Greenhouses Ltd., Vancouver, British Columbia, Canada.)

FIGURE 19.17 Caulking of gable ends. (Courtesy of Resh Greenhouses Ltd., Vancouver, British Columbia, Canada.)

panels, starting by pushing them under the ridge cap and bending them down over the purlins and attaching them to the sills (Figure 19.16). Use closure strips at the ridge cap and inside on the sills. Secure the panels to the sill with stainless steel wood screws having washers with neoprene seals underneath them. Screw into the wooden sill through the closure strip on the inside at 6″ spacing across the panels. With completion of the gable end panels caulk the joints between the gable ends and sidewalls using silicone or other outside weather-resistant caulking (Figure 19.17). The next steps include installing the glass and aluminum door, fan(s), and inlet shutter(s) for ventilation if doing this yourself and not a purchased pre-assembled gable. The completed greenhouse with hydroponic beds, fan, and inlet shutter is shown in Figure 19.18.

GREENHOUSE COMPONENTS

The greenhouse environment must be kept optimum to maximize plant yields. The two most important factors to control are temperature and light. Temperature control includes heating, cooling, ventilation, and air circulation. The heating requirements depend upon the crop grown, location of the greenhouse, and its type of construction materials. The nature of the covering material and its R-value determines the amount of insulation capacity of the structure. R-values were discussed earlier. The following example demonstrates the calculation of heat loss in a typical greenhouse.

FIGURE 19.18   Completed 8 ft × 12 ft greenhouse. (Courtesy of Resh Greenhouses Ltd., Vancouver, British Columbia, Canada.)

HEATING

The following example is to calculate the heat loss of a backyard greenhouse 10 ft wide (W) by 12 ft long (L) by 9 ft to the roof peak (H) having 6 ft high sidewalls (S). It is constructed of aluminum framing with 16 mm thick triple-wall polycarbonate cladding situated in a climate where the minimum extreme winter temperature would be about 0°F (−18°C). If under the extreme minimum temperature we wish to maintain 50°F (10°C), the temperature differential is 50 F (28°C) degrees (T). So, the heating system must be capable of raising the temperature of all the air volume within the greenhouse 50°F (28°C). The next step is to calculate the total exposed surface area (A) of the greenhouse. Refer to the diagram (Figure 19.19) for the dimensions of the greenhouse. The following websites give a simplified explanation for calculating heat loss for various shapes of greenhouses and have calculators that will do all the mathematics for you:

> www.gothicarchgreenhouse.com/heating.htm
> www.gothicarchgreenhouse.com/Greenhouse-Heater-Calculator.htm
> www.gothicarchgreenhouse.com/Greenhouse-Surface-Calculator.htm
> www.littlegreenhouse.com/heat-calc.shtml
> www.littlegreenhouse.com/area-calc.shtml
> www.sherrysgreenhouse.com/oldsite/GHheating.html

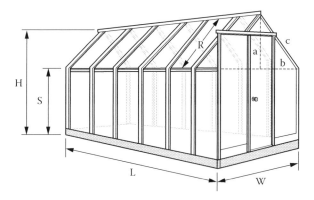

Length of hypotenuse (longest side) of triangle
pythagoras theorem: $a^2 + b^2 = c^2$

Roof width (R) = c

$$R = c = \sqrt{a^2 + b^2}$$

= Square root of the sum of the squares
of sides "a" and "b"

FIGURE 19.19 Calculation of gable greenhouse surface area. (Drawing courtesy of George Barile, Accurate Art, Inc., Holbrook, New York.)

Use the following equation:

1. Exposed Surface Area (A) = Area of Gables + Area of Walls + Area of Roof

Area of Two Gables: This is a traditional gable-shaped greenhouse. To calculate the area of the gables, we must calculate the bottom rectangular surface plus the gable-shaped component. For that calculate the area by use of triangles. Each upper portion of the gable is made up of two triangles (Figure 19.19).
The area of a triangle is: A = ½ base × height = ½ × 5 ft (triangle base is ½ of the gable width) × 3 ft = 7.5 sq ft
For the two triangles: 2 × 7.5 = 15 sq ft
Add the bottom portion: 10 ft × 6 ft = 60 sq ft
For two gables: 2 × (15 + 60) = 150 sq ft
Area of Walls (2): 2 × (6 ft × 12 ft) = 144 sq ft

Area of Roofs (2): For this we use the triangle to find the length of the hypotenuse (longest side) using the Pythagoras theorem equation: $a^2 + b^2 = c^2$; where c = the length of the hypotenuse (see diagram in Figure 19.19).
$c^2 = a^2 + b^2 \Rightarrow c^2 = 3^2 + 5^2 = 9 + 25 = 34$: Therefore, c is the square root of 34: 5.8 ft. You may use a calculator to get the square root of a number. The longest side is 5.8 ft, which is the end width of the roof (R).
The total roof area is (two sides): $2 \times L \times R = 12$ ft $\times 5.8$ ft = 139 sq ft
Total Exposed Surface Area (A): 150 sq ft + 144 sq ft + 139 sq ft = 433 sq ft

2. Determine the greenhouse construction factor (C):

C = 1/R-value
The R-values were given in the table earlier for the various covering materials. For 16 mm triple polycarbonate R = 2.5.
    Therefore, C = 1/2.50 = 0.40

3. Determine the wind correction factor (W): Refer to the following table.

Wind Velocity (mph)	"W" Factor
15 mph or less	1.00
20 mph	1.04
25 mph	1.08
30 mph	1.12
35 mph	1.16

For 30 mph use: W = 1.12

4. Determine heat loss in British Thermal Units (BTU):

$$BTU = T \times A \times C \times W = 50 \times 433 \times 0.40 \times 1.12 = 9700 \text{ BTU}$$

There are BTU calculators on the Internet websites listed earlier. You can simply plug in the dimensions of the greenhouse, the R-value for the covering, and the temperature differential and they will calculate the total BTU loss.
To determine the projected electrical cost for heating convert the BTU to Watts as follows:

Watts/h = BTU/3.413. Therefore: 9700 BTU/3.413 = 2842 watts or 2.84 KWH (kilowatts/hour)
Cost of operating under the most extreme cold conditions (assume cost of power is $0.10/KWH):
2.84 KWH × $0.10 = $0.28

However, note that this would be under the most extreme minimum temperature. Normally, this period would be only a few hours during the coldest day of the year. To project an annual cost of heating the greenhouse over a year, you will need to visit the website of the National Weather Service and look up the weather data for your specific location. The website is www.ncdc.noaa.gov (National Climatic Data Center). Click on "Most Popular Data" and go to "4. 1971-2000 US Climatic

Normals Products;" then go to "1971–2000 Normals Products Page" and finally to "Monthly Station Climate Summaries (CLIM20)." Then type in your state and city for your area records.

While you must design your heating requirements on the minimum extreme temperatures, when calculating the monthly predicted costs adjust the calculations according to the climate data of your area. The following example is for Seattle-Tacoma (airport), Washington (Climatic Div. WA 3) from the weather data charts (Table 19.2). The minimum extreme (using records from 1971 to 2000) for January was 0°F (−18°C), the minimum monthly mean extreme temperature was 34.8°F (1.6°C), and the daily minimum for January was 35.9°F (2.2°C) over the period.

To calculate the projected heating cost, use the daily mean for each month and add all of the months or use the annual mean and the temperature differential needed to heat the greenhouse to 70°F (21°C) (65°F or 18°C night, 75°F or 24°C day). For example, for January using the daily mean of 41°F (5°C) the temperature difference is 70°F–41°F = 29 F° (16 C°).

Then, multiply the previously calculated BTU by the fraction of monthly temperature difference divided by the extreme temperature difference.

For example, for January: 9700 BTUH × 29/50 = 5626 BTUH

The cost of electricity for the month of January: 5626/3.413 = 1648 Watts/h or 1.65 KWH

For 24 h: 24 × 1.65 = 39.6 KWH or about 40 KWH

Cost per day at $0.10/KWH: 40 × $0.10 = $4.00

Total cost for the month of January: 31 × $4.00 = $124.00

Do this for the rest of the months of the year or use the annual daily mean (52.3°F) as follows:

TABLE 19.2
Recorded Temperatures over the Period 1971–2000 for Seattle, Washington

Month	Min. Extreme (°F)	Min. Mean Extreme (°F)	Daily Min. (°F)	Daily Mean (°F)	Extreme Max. (°F)	Daily Max. (°F)
Jan.	0	34.8	35.9	40.9	64	45.8
Feb.	1	35.9	37.2	43.3	70	49.5
March	11	41.3	39.1	46.2	75	53.2
April	29	45.8	42.1	50.2	85	58.2
May	28	51.8	47.2	55.8	93	64.4
June	38	56.4	51.7	60.7	96	69.6
July	43	61.2	55.3	65.3	100	75.3
Aug.	44	61.9	55.7	65.6	99	75.6
Sept.	35	55.9	51.9	61.1	98	70.2
Oct.	28	49.7	45.7	52.7	89	59.7
Nov.	6	35.8	39.9	45.2	74	50.5
Dec.	6	35.3	35.9	40.7	64	45.5
Annual	0	34.8	44.8	52.3	100	59.8

Factor to multiply the extreme temperature BTUH by is:

Temperature difference of 70°F–52.3°F = 17.7 F degrees.
9700 BTUH × 17.7/50 = 3434 BTUH
WH is 3434/3.413 = 1006 watts/h or 1 KWH
KWH per day: 24 × 1 = 24 KWH
Cost per day: 24 KWH × $0.10 = $2.40
Annual Projected Heating Cost: 365 × $2.40 = $876.00

VENTILATION

Solar radiation is the force causing transpiration in plants. As the sunlight shines on the leaves of the plants, their temperature rises causing the stomates to open and transpiration to occur. As the water evaporates from the leaves the leaves are cooled.

Only about 70% of the light passes through the greenhouse roof cover, the other 30% is reflected. Plants absorb about 70% of this available light and re-radiate the other 30%. Therefore, plants absorb about 50% (70% of 70%) of the incoming sunlight using this for transpiration as shown in Figure 19.20. With a mature large crop, only 20% (30% of 70%) of the solar energy is left to heat the air. When there is no crop in the greenhouse, about 70% of the available total solar radiation is available to heat the greenhouse air.

During high solar light from spring through early fall, temperatures will rise in the greenhouse due to the "greenhouse effect." This effect is from the wavelength of the light changing as it enters the greenhouse through the covering and the heat component cannot escape. As a result, the air temperature in the greenhouse rises.

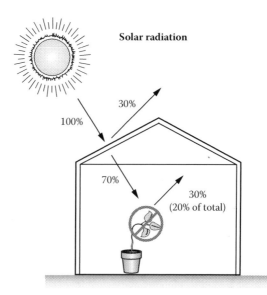

FIGURE 19.20 Solar radiation. (Drawing courtesy of George Barile, Accurate Art, Inc., Holbrook, New York.)

It soon exceeds the optimum maximum temperature range for the plants causing the stomates to close and restrict photosynthesis, which reduces growth and yields. To reduce excessive heat build-up, the air in the greenhouse must be forced out and be replaced by cooler outside air. Natural ventilation with roof vents will assist in the air exchange. This is the least expensive method of cooling as the vents can be operated without electricity using solar vent openers. A black cylinder filled with paraffin wax expands with heat and pushes a piston that opens the vents automatically with metal arm linkage as the temperature changes (Figure 19.21). These vent openers can operate roof and side vents. There are a number of different sizes of vent openers (about $50) that are capable of opening vents 12–15″ with vents weighing from 15 to 25 lbs and more. Operating temperatures to start opening can be adjusted from 60°F to 78°F (15.5–25.5°C) and maximum opening at 86–90°F (30–32°C).

Side louvers to permit cool fresh air to enter near the base of the greenhouse can also be operated by the automatic solar vent openers. With installation of both roof vents and inlet louvers, the movement of air will flow from the lower cooler air entering below and rising as it is heated within the greenhouse to escape through the roof vents (Figure 19.22). Wind passing over the open roof vents also creates a pressure difference that acts like a vacuum and sucks the heated air out of the vents. This movement of air with the temperature gradients exchanges the air within the greenhouse adding carbon dioxide as well as maintaining optimum temperatures.

If you live in a region where summer temperatures are very high, you may install an exhaust fan and inlet shutters to force the air out of the greenhouse (Figure 19.23). This forced ventilation should exchange the entire air volume of the greenhouse within 1 min to minimize temperature gradients from one end to the other. If the

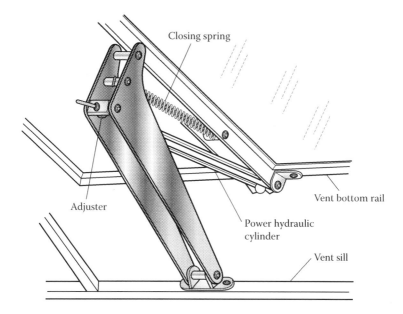

FIGURE 19.21 Roof vent and opener. (Drawing courtesy of George Barile, Accurate Art, Inc., Holbrook, New York.)

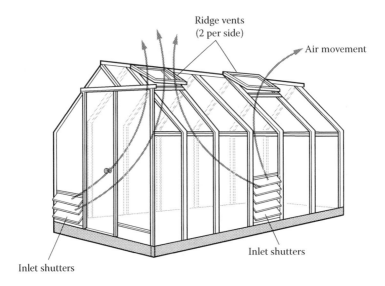

FIGURE 19.22 Natural air circulation within a greenhouse. (Drawing courtesy of George Barile, Accurate Art, Inc., Holbrook, New York.)

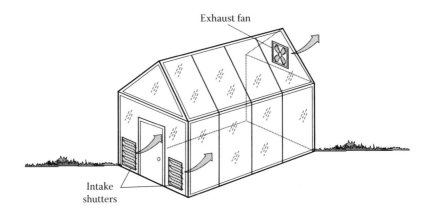

FIGURE 19.23 Forced ventilation with exhaust fan and intake shutters. (Drawing courtesy of George Barile, Accurate Art, Inc., Holbrook, New York.)

exhaust fan is operating, roof vents must be closed to prevent the air from short circuiting down from the roof vents directly outside the fan without being forced through the crop. To determine the size of the exhaust fan and shutters needed, you must calculate the total air volume of the greenhouse and express it as cubic feet of air per minute (CFM). The total volume of air is the width × length × height of the greenhouse. The industry uses a factor of 10 ft times the width × length or 12 ft times the width × length for more southerly locations. Using our example earlier for heating, the greenhouse was 10 ft × 12 ft so the total CFM calculation is: 10 ft × 12 ft × 12 ft = 1440 CFM. Or we can take the total area of the gable end of the greenhouse and multiply it by the length (12 ft) as follows: The gable end area was 150 sq ft. Multiply that by the 12-ft length: $150 \times 12 = 1800$ CFM.

Now refer to charts of various sizes of fans and shutters to determine the model of exhaust fan needed. Then, match the total CFM output of your selected exhaust fan to the CFM ratings for the shutters. There will be one exhaust fan and two inlet shutters. Shutters are square in dimension starting at the smallest ones 12″. The next are 16″, 18″, 20″, 24″, and up to 54″. A 16″ shutter has a capacity of close to 900 CFM, so two of those would be adequate. The next size, 18″, is rated at 1125 CFM. When purchasing an exhaust fan, it is best to have a two-speed motor activated by a two-stage thermostat so that as the temperature rises in the greenhouse the first thermostat setting operates the slower motor speed and as temperature continues to rise the second setting would initiate the faster motor speed. Set the two ranges at least 5°F (2.8°C) apart.

Another method of cooling is to prevent some of the direct sunlight from entering the greenhouse by use of a shade cloth over the top of the greenhouse. Usually, 35% shade is adequate. This would only be used during the hottest time of the year when most sunlight occurs. You can also white wash the outside of the covering with a special white paint that can be easily removed by washing. It generally has a life expectancy of 3–4 months so that it is easily removed by late fall when the sunlight is no longer intense and when you then need as much light as possible as the season progresses into winter.

Evaporative cooling pads are also effective in lowering very high temperatures under low ambient relative humidity (RH). Evaporative cooling pads would be positioned on one end of the greenhouse opposite the exhaust fans. In that case you would need two exhaust fans near the door end and have the cooling pad on the opposite gable end. Still another system is to use high-pressure fogging. The fogging misters are mounted within the greenhouse. Their efficiency, like the evaporative pads, depends upon low ambient RH. I do not recommend these evaporative cooling systems for small greenhouses as they are costly.

An article on greenhouse ventilation by the University of Connecticut states that fan ventilation can consume from 0.5 to 1 kilowatt hour (KWH) per square foot of greenhouse area per year. So the potential annual cost for the 120 sq ft hobby greenhouse example would be: 120 KWH $\times$ \$0.10 = \$12.00.

For more accurate calculation of a given location, use the number of days of sunshine obtained from the weather data. During those sunny days use about 8 h per day as that is when the light would be most intense. This would be only for the months from May through September when day length is longest and temperatures highest.

For our example greenhouse, a 16–18″ exhaust fan would give adequate ventilation of 1200 CFM. This type of fan has a 1/10th HP (horsepower) motor that uses 110/120 volts (V) and draws 1.5 amperes (I) of current. The power law is: P (watts) = Voltage (V) $\times$ Current (I). Therefore, the power utilized per hour of fan operation is: $P = 110 \times 1.5 = 165$ watts per hour.

To operate this fan for 8 hours it would consume: 8×165 watts = 1320 watts or 1.32 kilowatts per day or 1.32 KWH per day. If the cost of power is \$0.10 per KWH, the daily cost would be $1.32 \times \$0.10 = \$0.13$. If the number of full sunlight days is 20 days per month over this period May through September (5 months), the total power consumption for the period would be 5 months $\times$ 20 days $\times$ 1.32 KWH = 132 KWH. The cost based on \$0.10 per KWH would be \$0.10 $\times$ 132 KWH = \$13.20.

FIGURE 19.24 Evaporative cooler unit. (Drawing courtesy of George Barile, Accurate Art, Inc., Holbrook, New York.)

A more effective cooling system for a backyard greenhouse is a greenhouse evaporative swamp cooler (Figure 19.24). This combines the exhaust fan and cooling pad within one unit. It draws hot outside air through evaporative cooling pads within the unit using a blower. As the water evaporates it takes the heat out of the air, resulting in pushing cool air into the greenhouse. This is a positive pressure system whereby the cooler pushes air into the greenhouse and allows it to exit through roof vents or exhaust shutters (Figures 19.25 and 19.26). When using an exhaust fan, it sucks air into the greenhouse through inlet vents and expels the hot air out. The positive pressure method is much superior as it also assists in preventing insects from entering the greenhouse.

Note that the evaporative cooler is mounted close to the base of the greenhouse to bring in the coolest air possible and the exhaust shutters are mounted high near the height of the eaves to expel hot air (Figure 19.26). The exhaust fan, on the other hand, is installed high near the eaves height (Figure 19.23) with its intake shutters mounted low within a few feet of the ground to bring in cool air. The swamp cooler pushes out heated air via the exhaust shutters or roof vents (Figures 19.25 and 19.26) while the exhaust fan sucks in cool air at the base of the greenhouse and exhausts it higher at the eaves height (Figure 19.23).

The amount of cooling provided through evaporation is a function of the ambient RH. The lower the outside RH, the more cooling capacity is available by evaporation. When the RH is lowest during the hottest times of the day, the air temperatures can be reduced significantly within the greenhouse. For example, if the outside temperature is 88°F (31°C) and the RH is 54%, the cooled air temperature entering the greenhouse

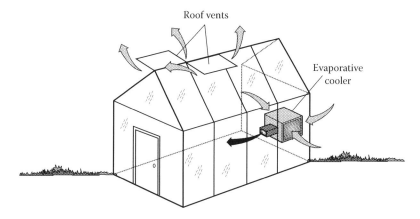

FIGURE 19.25   Evaporative cooler with roof ridge vents. (Drawing courtesy of George Barile, Accurate Art, Inc., Holbrook, New York.)

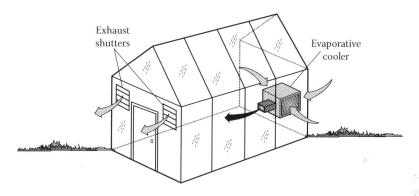

FIGURE 19.26   Evaporative (swamp) cooler with exhaust shutters. (Drawing courtesy of George Barile, Accurate Art, Inc., Holbrook, New York.)

will be 78°F (25.5°C). Similarly, with an outside temperature of 106°F (41°C), RH of 38% the air temperature after cooling would be 78°F (25.5°C). However, with a combination of high temperatures and RH, the cooling capability of the evaporative cooler is reduced since little water will evaporate as the ambient RH approaches 85%.

The size of evaporative cooler needed for the example hobby greenhouse is determined by the total CFM as for the earlier exhaust fan. The evaporative cooler comes with a two-speed fan, and can hence service from 1400 to 2800 CFM. A model to satisfy this need has one-eighth HP, is 110/120 volts, and draws 5.4 amperes. As a result, this unit will consume more power than an exhaust fan system. The power consumption is: $P = VI$; $P = 110 \times 5.4 = 594$ watts. Cost per 8 h of operation is $594 \times 8 = 4752$ watts or 4.75 KW. That is 4.75 KWH per day. For the 5-month cooling period of the year, the total power usage is 100 sunny days × 4.75 KW = 475 KWH. The cost would be $0.10 × 475 = $47.50.

The intake shutters or exhaust shutters can operate on the pressure differences from the fans or they may be equipped with a small motor to open and close them in coordination with the operation of the exhaust fan or evaporative cooler blower. The shutter motors are very small drawing about 0.17 amperes at 120 volts for some models. The operational cost of two shutters is much less than the exhaust fan. Projected power and costs are as follows: $P = VI$; $P = 120 \times 0.17 = 20$ watts. For two shutters that is 40 watts. Total power consumption for the 5-month period is $100 \times 8 \times 40 = 32$ KWH. The cost would be $\$0.10 \times 16 = \$3.20$ for the season.

TOTAL ESTIMATED HEATING AND COOLING COSTS

Heating: $876.00
Cooling: $13.20 (for an exhaust fan and automatic shutters not motorized)
Cooling: $47.50 (for an evaporative cooler and non-motorized shutters)
Motorized shutters: $3.20

The total cost is either $926.70 for an evaporative cooler with motorized shutters or $889.20 for an exhaust fan with automatic shutters (not motorized). Even these two systems are not significantly different in the annual cost for temperature control.

POTENTIAL ANNUAL PRODUCTION

Overall, the cost of heating and cooling would be covered by the value of the vegetables. The expected annual yields from the greenhouse production of vegetables are dependent upon the amount of sunlight the plants will receive. The following yields are based upon commercial greenhouse production in the Vancouver, BC area, which is very similar to Seattle, WA, the location of our earlier example. Table 19.3 gives annual production per crop and in the last two columns are projections for a combination of crops in a 10 ft × 12 ft backyard greenhouse. Herbs

TABLE 19.3
Potential Annual Production in a 10 ft × 12 ft Greenhouse

Crop	Area/Plant (Plt)	Crop Life (Days)	Crops/Year	Annual Pdn/Plt	No. of Plts	Annual Yield (lbs)
Tomatoes	3.5 sq ft	9 months	1	40 lbs	12	480 lbs
Peppers	3.5 sq ft	9 mo.	1	30 lbs	8	240 lbs
European Cucumbers	9 sq ft	3–4 mo.	3	100 lbs	2	200 lbs
Eggplants	3.5 sq ft	3–4 mo.	3	20 lbs	2	40 lbs
Lettuce (Bibb or Leaf) (Raft)	4 plts/sq ft	30 days	11–12	11–12 Head/Yr.	48	528–576 head
Herbs (Plant Towers-PT)	44 plts/10 sq ft	9 mo. (3wks/cut)	1 crop (12 cuts/year)	¼ lb/year (depending on type of herb)	44	11 lbs
Arugula, Basil (PT/Raft)	4 plts/sq ft (Raft)	30 days	11–12	11–12 head	—	—
	44 plts (PT)	3 mo.	4	4 oz/crop/plt	44	44 lbs

should be grown in plant towers to increase the production per square foot of greenhouse area.

TYPICAL CROP LAYOUT PLAN (10 FT × 12 FT GREENHOUSE)

In a 10 ft wide by 12 ft long greenhouse, there will be three beds. There is one on each side and one in the center of the greenhouse. The two side beds are 12 ft long and the center one is 9 ft long to permit entrance from the door that swings outward. A 2-ft wide aisle is allowed for access between the beds. Vine crops are supported vertically from support cables attached by eye hooks and turn buckles along the roof frame. Bolt the eye hooks and turnbuckles into the aluminum ribs of the greenhouse. Plastic vine twine supports the vine crops from the overhead cable with special hooks with additional string called "Tomahooks."

In a 10 ft × 12 ft greenhouse, we may grow 12 tomatoes on one side in either bato buckets or rockwool/coco coir slabs. Details of the hydroponic crops and their growing systems are given in Chapter 20. The central bed contains two European cucumbers, eight peppers, and two eggplants. Due to the large size of eggplants, grow one plant per pot. The other area of 2 ft × 12 ft has two plant towers with herbs in the one, and basil and arugula in the second one. The remaining 2 ft × 6 ft is for a floating raft-culture system. It is used for lettuce. The raft system can fit 48 head of lettuce. Lettuce, arugula, and basil may be grown in either NFT or raft culture. Alternatively to the plant tower, the arugula and basil could be grown in the raft culture system in place of some lettuce. The production for the greenhouse is summarized in Table 19.3. The projections use plant towers for herbs, basil, and arugula. A plan for these crops is given in the diagram (Figure 19.27).

CONTROL PANEL

Cut and paint a piece of ¾″ thick plywood for a backing for all of the controls as shown in the diagram (Figure 19.28). The length must be sufficient to span across two of the greenhouse vertical frame bars along the back end wall near the nutrient tank. Make it at least 2 ft wide to fit the components. Locate it opposite one of the aisles for easy access and at a height just above the irrigation header. The electrical cable from your residence enters at the back wall of the greenhouse behind the nutrient tank and is connected to a breaker panel. The breaker needs a 220 volt, 30-amp circuit for the electric heater, four 110 volt, 15-amp circuits for outlet socket boxes, and two time-clocks. From the breaker panel, using electrical conduit or other approved water-proof cable, place three dual socket outlets for the exhaust fan, and three 8-ft, dual tube, high output fluorescent lights. One time-clock operates the lights and the other time-clock is wired directly to another circuit that has a dual outlet box for the pump. The outlet for the 220-volt heater is wired directly into the 220 volt, 30-amp circuit of the breaker panel. The heater has its own built-in thermostat. The exhaust fan is operated by a thermostat hung in the middle of the greenhouse. The control panel will keep all the wiring neat and all controls centralized for easy operation. This arrangement of components for a backyard greenhouse is shown in Figure 19.29.

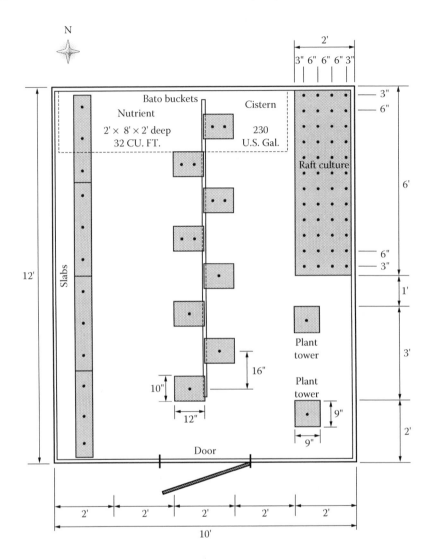

FIGURE 19.27 Crop plan for a backyard greenhouse with hydroponic systems. (Drawing courtesy of George Barile, Accurate Art, Inc., Holbrook, New York.)

As mentioned earlier under "Site Preparation," lay the approved underground cable before starting the preparation of the greenhouse base. If a propane or natural gas heater is used instead of the electrical heater, the electrical circuit for the greenhouse can be reduced to 60 amps. If natural gas is to be used, it will have to be installed according to local codes by an approved gas installation company. The trench for the placement of the electrical line may be able to also contain a water line for the greenhouse. Again, local codes dictate the depth of these trenches and what utilities may be placed together. A water line of ¾″ black poly tubing approved

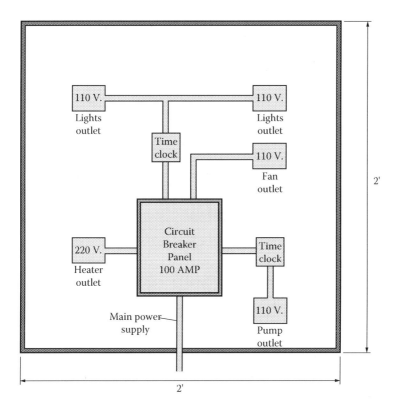

FIGURE 19.28 Control panel layout. (Drawing courtesy of George Barile, Accurate Art, Inc., Holbrook, New York.)

for the pressure of your water source would be adequate for the greenhouse usage. Natural gas heating would be the most efficient and lowest cost form of heating.

PROJECTED ANNUAL REVENUES

Combining the projected annual production in Table 19.3 with an expected price to purchase the product in a supermarket, the overall revenues are given in Table 19.4.

The revenues, of course, make the assumption that you would consume that amount of salad crops a year. Most of these products could be consumed by a family of four to six people providing they are much focused on eating vegetables as an important part of their diet. The lettuce is more than would be needed, so the arugula and basils could be grown in the raft system with the lettuce and the second plant tower freed up for more herbs, bok choy, chard, spinach, and even some flowers such as viola, nasturtiums, petunias, marigolds, and so on, that normally do well in hanging baskets. The bok choy and chard could also be grown in the raft system replacing some of the lettuce. If this is still too much produce, share it with relatives and/or neighbors. You might even sell some to them at a discounted price from that of the supermarket.

FIGURE 19.29 Greenhouse with hydroponic and environmental control components. (Courtesy of Resh Greenhouses Ltd., Vancouver, British Columbia, Canada.)

TABLE 19.4
Projected Annual Revenues in a 10 ft × 12 ft Greenhouse

Crop	Annual Yield	Weekly Production	Unit Price	Annual Revenues ($)
Tomatoes	480 lbs	9.2 lbs	$2.00	$960
Peppers	240 lbs	4.6 lbs	$4.00	$960
European Cucumbers	200 lbs	3.8 lbs	$1.50	$300
Eggplants	40 lbs	0.77 lbs	$1.50	$60
Lettuce	550 head	11 head	$1.50	$825
Herbs	11 lbs	3.4 oz	$20.00	$220
Arugula, Basil	44 lbs	6.8 oz	$10.00	$440
Total				$3765

20 Hydroponic Systems for Backyard Greenhouses

INTRODUCTION

Most hydroponic systems are adaptable to small-scale backyard greenhouses. However, not all are practical. The choice also depends upon the crop grown. Lettuce, arugula, bok choy, basil, and some herbs do best in a nutrient film technique (NFT) or raft culture system. Vine crops prefer pots or slabs of perlite, coco coir, rockwool, or mixtures of these substrates. To increase the production of low-profile herbs, bok choy, and strawberries, plant towers are the preferred system using peat, perlite, coco coir, or mixtures of these media. Usually in a backyard greenhouse, we wish to grow most of these crops together. The first step then is to determine how much of each crop you like in your salads on a weekly or monthly basis. Next, is to decide on the best system to use for each crop and how much area is to be occupied in the greenhouse by each with its hydroponic system. Finally, make a detailed plan of the location of each crop in the greenhouse and the specific area occupied by each one.

Hobby greenhouses come in many dimensions as was described in Chapter 19. The width limits the number of plant rows or beds that it can contain. Greenhouses of 8-ft width will fit two beds, and 10–12-ft wide greenhouses have three beds. The center of the greenhouse is 9-ft to 10-ft high from the base. If the structure has the base (sills) set on a concrete block foundation, the height may be raised several more feet. This extra height helps greatly to accommodate tall vine crops. Locate European cucumbers, peppers, and eggplants in the center bed of the greenhouse where the height is greatest, as these plants are difficult to lower. Tomatoes can be located on one of the side beds and low-profile plants and plant towers on the other side and as shown in the crop plan of the last chapter (Figure 19.27).

HYDROPONIC SYSTEMS

SLABS

Pre-wrapped rockwool and coco coir slabs are available at hydroponic shops or online. They come in widths of 6″, 8″, and 12″ by 3″ thick by 3 ft long. They cost $6–7 each. I highly recommend purchasing these slabs instead of making them from polyethylene. Especially the coco coir slabs as they will be leached to remove any sodium chloride from the substrate. Rockwool slabs are only available as wrapped. The slabs are roughly 3 ft long. Four slabs will make up a 12-ft row to fit in a hobby greenhouse of 12-ft length. If you wish to grow all vine crops using only slabs, four will fit on each side and three in the center row of the greenhouse. The slabs are perfectly suitable for low-profile plants. Simply make the plant holes at 6″ centers within the slabs and place three rows of slabs (using 6″ wide slabs) 3″ apart to get a 24″ growing bed. The disadvantage of using the slabs instead of raft or NFT for low-profile plants is the high cost of the slabs. To obtain a growing area of 2 ft wide by 9 ft long a total of 3 × 3 = 9 slabs would be needed. The cost would be approximately $60. In addition, it is difficult to re-use the slabs for more than two to three crops as they may become contaminated and/or get damaged physically during the harvesting of the plants. That would result in replacing the slabs 5–6 times a year so the cost could escalate to $300–350 annually. For these reasons, stay with the conventional methods of growing specific crops.

For tomatoes, peppers, and eggplants locate three plants per slab (Figure 20.1). That is 12 plants per row of four slabs. Space the plants within a slab at 6″ from each end with one in the middle. For European cucumbers transplant two plants per slab, one at 9″ from each end of the slab and space the slabs 6″ apart end-to-end. Then, in a 12-ft row there would be three slabs instead of four slabs as with the tomatoes, peppers, or eggplants. That would give six cucumber plants per row.

The slabs are set on top of a 1–2″ thick Styrofoam sheet to insulate the roots from the cold floor of the greenhouse (Figure 20.2). Alternatively, construct raised beds to set them on, but that will take away from the overall height to train the plants. Cover the Styrofoam insulation with black polyethylene plastic to permit drainage

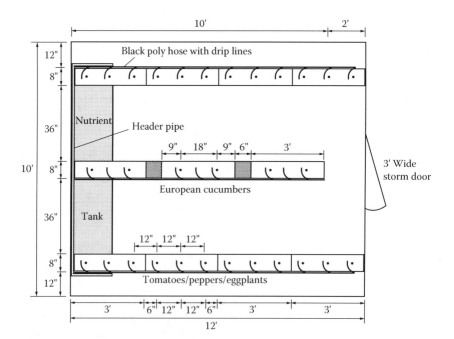

FIGURE 20.1 Plant spacing in slabs of various vine crops in a 10 ft × 12 ft greenhouse. (Drawing courtesy of George Barile, Accurate Art, Inc., Holbrook, New York.)

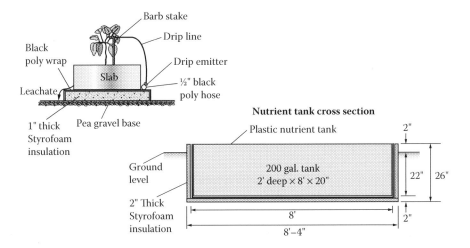

FIGURE 20.2 Open system of slabs on Styrofoam insulation and sunken nutrient tank. (Drawing courtesy of George Barile, Accurate Art, Inc., Holbrook, New York.)

to occur over the sides of the Styrofoam to minimize the growth of algae. This is an open system as shown in the diagram. To construct a re-circulation system to collect the leachate and return it to the cistern tank, use rigid plastic channels. These channels are available from hydroponic shops or online. Place the channel on top of the Styrofoam insulation, sloping it about 4% back to the cistern.

You could also construct beds from plywood and a steel-frame. Make low beds to get a slope of 3% back to the cistern. That is a 4–5″ slope back to the top of the cistern. The bed would be 6″ high at the entrance end of the plant row sloping to the cistern having 2″ of freeboard (above ground level). Make the bed with 2″ sides using 1″ × 2″ lumber. Construct the width 2″ wider than the slab width. Using 8″ wide slabs make the width of the bed 10″ (inside width). Cut the Styrofoam the same width as the slabs as they will sit directly on top of the Styrofoam. Staple 6-mil thick black polyethylene at the top edge of one side of the bed, bring it over the Styrofoam and push it between the Styrofoam and the other edge of the bed to form a channel where the leachate can run along and enter the cistern at the lower end as shown in Figure 13.22. Staple the other side of the black polyethylene liner to the upper edge of the bed side.

Bato Buckets

Bato buckets are special pots designed in Holland for coarse media such as perlite, expanded clay rocks, or pea gravel. With these coarse substrates, the bato bucket maintains about 2″ of solution at the bottom. This level of solution is regulated by a siphon pipe. The bato buckets sit on top of a 1½″ diameter PVC drain pipe. Holes of 1″ diameter are drilled into the top of the drain pipe to fit the siphon elbow from the base of the bato bucket. The holes are spaced at 16″ centers along the drain pipe. Begin the first hole about 8″ from the end of the drain pipe to allow a cap at the end of the drain pipe. Do not glue the cap as that is access to clean out the drain pipe. The drain pipe lies on the greenhouse floor with a very slight slope back to the cistern by adding some pea gravel under the pipe and pots. Bato buckets cost about $6.50 each and can be re-used between crops for at least 5 years. Clean them between crops with a 10% bleach solution by soaking them for 1 h or slightly longer.

Eight bato buckets in 10–12 ft will give adequate plant spacing (Figure 19.27). Each bucket will hold one European cucumber or two of tomatoes, peppers, or eggplants. Usually it is better to plant one eggplant per bucket as eggplants have large leaves that intercept more light than tomatoes or peppers. Other low-profile crops such as herbs, lettuce, arugula, basil, bok choy, cabbage, cauliflower, broccoli, green onions, bush beans, and many more can be grown in bato buckets. Between crops simply sterilize the pots and replace new substrate. They are more economical than slabs to grow these crops since only the substrate needs replacement between cropping cycles. Herbs, of course, can be harvested for up to 10 months, depending upon the herb's growth cycle. Even strawberries would grow in bato buckets. During the hot weather seasons from late spring to early fall, grow cool-season crops such as cabbage, cauliflower, and broccoli outside in your soil garden as the greenhouse would be too hot for them at that time of the year.

Plant Towers

Plant towers increase the number of plants that can be grown in a unit area of the greenhouse compared to ground-level beds. Usually, at least 6 times the production is achieved in plant towers compared to ground beds. Other advantages include easy maintenance and harvesting of the plants, clean product as the crop is above the ground level, and the solution can be re-cycled. The dimensions of the Styrofoam pots for the plant towers are 9″ × 9″ × 8″ tall. Each pot is specially designed so that the pots sit one on top of the other without nesting by fitting their bases into four cuts in the lip of the pot below. This is a patent design. The pots are sold by Verti-Gro, Inc. in Florida (see Appendix) at a price of $4.50 each. They will easily last 5 years. The bottom pot should sit on top of a collection pot that returns the solution back to the cistern via a 1½″ drain pipe. Stack up to seven pots or more depending upon the greenhouse roof height. A galvanized electrician's conduit with a 1″ diameter PVC sleeve supports the pots vertically. Secure a support cable on the ribs of the greenhouse centered directly above the row of plant towers. The conduit is attached to the cable to keep the plant towers vertical. The remaining details of setting up the plant towers were given in Chapter 13, on "Plant Towers" including Figures 13.26 and 13.27. For only two plant towers as in Figure 19.27 of the greenhouse crop layout, let the plant towers drain to waste.

Plant towers are best for herbs, bok choy, and flowers such as marigolds, nasturtiums, viola, petunias, and all hanging-basket type of flowers. Do not use them for lettuce as harvesting often and changing the plants is a lot of work, nonetheless, the towers will produce very nice lettuce. Do not grow any vine-crops or even bush-type tomatoes, peppers, eggplants, and so on as these plants cast a lot of shade on the ones immediately underneath causing very poor yields in the lower plants. A detailed description of the irrigation system follows under that section of this chapter.

RAFT CULTURE

This hydroponic system is best for lettuce, basil, arugula, and some herbs. It can easily be set up in a backyard greenhouse. In our example, a raft system could occupy the bed area along one side or in the center as a full bed length or a portion of it. The first step is to decide on how many lettuce, basil, arugula, and some herbs are eaten weekly. From this extrapolate, the area of raft culture needed to fulfill your requirements assuming the cropping cycle for these plants is about 6 weeks. Spacing of these plants is 6″ × 6″ so you will get four plants per square foot of the bed surface area. As shown earlier in the plant layout (Figure 19.27), 48 plants will fit in a bed 2 ft wide by 6 ft long. The raft system is self-contained so its operation is independent of the other hydroponic systems in the greenhouse. It will have an air pump above the bed that circulates air by a poly hose to air stones in the pond. Locate the air pump above the top level of the pond in case it should stop and thus not allow solution to flow back to the pump. These components and supplies are available from an aquarium store or Aquatic Eco-Systems, Inc. in Florida (see Appendix).

In northerly climates with cold soils, especially during the winter months, low temperatures will chill the nutrient solution below optimum of the raft system, which is about 65–68°F. Even in the greenhouse with a weed mat and gravel base as described in the preparation of the site for the greenhouse in Chapter 19, the cold will affect the temperature of the nutrient solution. For this reason, place a 2″ thick Styrofoam board under the area where the raft pond is to be built. During long-term cold periods, the solution may have to be heated with an immersion heater.

Construct the sides of the pond with 2″ × 10″ treated lumber. Set these sides on top of the Styrofoam. In constructing one of the side beds of the greenhouse, make the inside dimensions 48½ wide by a length 4″ less than the inside length of the greenhouse. When cutting the Styrofoam boards (rafts) allow at least ½″ play (less length than the inside length of the bed). Screw the lumber together with stainless steel screws and glue the joints. Set the perimeter frame on the Styrofoam and then install a 20-mil swimming pool liner, folding the corners similar to making a parcel followed by nailing the top edge onto the wood frame using a cedar wood lathe or aluminum angle as shown in Figure 13.2.

The Styrofoam boards can be either 2 ft × 4 ft or 4 ft × 4 ft in dimensions. Make the holes for the transplant plugs at 6″ × 6″ spacing starting 3″ from the edge of the boards as shown in Figure 13.4. Each 4 ft × 4 ft board should hold 64 lettuce, arugula, or basil plants. It is easier to handle 2 ft × 4 ft boards especially when harvesting as the plants weigh up to 6–8 ounces or more each. With fewer plants, it is unlikely that the boards will break from the weight as occurs with 4 ft × 4 ft boards.

EBB AND FLOW POT SYSTEM

This system was discussed in Chapter 13. Several types of these systems are available commercially on the Internet or at hydroponic shops (see Appendix). Some of the systems use a substrate such as expanded clay aggregate, coco coir, perlite, granular rockwool, and mixtures of coco coir, perlite, and rice hulls combined at different ratios. The expanded clay aggregate gives best drainage, is inert and pH neutral. It is re-usable by cleaning and sterilizing it between crops using a 10% bleach solution or hydrogen peroxide followed by rinsing. However, with subsequent cropping roots may enter into the clay particles and make it difficult to sterilize. This substrate can easily be replaced with new between crops.

Most of the ebb and flow systems use 5-gal buckets or some form of growing pots of that volume. The pots have a felt or screen liner bag that contains the substrate to prevent clogging of the drain outlet as particles may come off the substrate with continued irrigation and drainage cycles during the cropping period.

The components of the system include a large nutrient solution reservoir of 50 gallons or larger. A submersible pump operated by a time-clock that pumps water from the storage tank to a smaller distribution reservoir of about 5–10 gal is shown in Figure 20.3. The inlet line is attached to a float valve (also acts as a safety backup) that regulates the inflow of solution to the distribution tank. From the distribution tank, a ½″ black poly tubing connects to each grow pot. As the solution flows to the distribution tank, it also flows by gravity to each grow pot filling it from the bottom. There are basically two systems, one the fill-up system and the other the drainage system (Figure 20.4). During the fill cycle, a timer activates the pump in the large reservoir and the solution flows to the distribution tank. On the return cycle, a second time-clock activates the pump in the distribution tank to pump the solution back to the main reservoir as the first timer shuts off the pumping cycle to the distribution tank. The two timers are synchronized so that as one is operating the other is off. The solution flows back from the grow pots to the distribution tank where it is pumped back to the solution reservoir. A second float switch on the bottom of the distribution tank will sense when the growing pots are empty to stop the drain pump from operating even if the return cycle timer is still activated. Ebb and flow cycles are timed according to the plant water needs. The fill and drain cycle levels may be regulated by electronic sensor switches that operate the pumps via a controller. This system was also discussed earlier in Chapter 13.

Modular ebb and flow system plan view

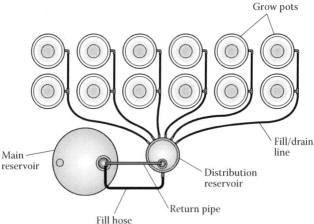

FIGURE 20.3 Ebb and flow plan of 12-pot setup. (Drawing courtesy of George Barile, Accurate Art, Inc., Holbrook, New York.)

Ebb and flow pot system fill cycle

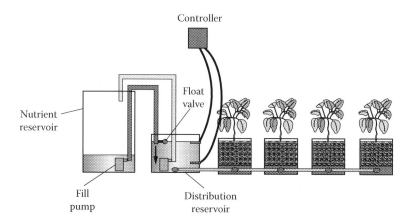

Ebb and flow pot system drain cycle

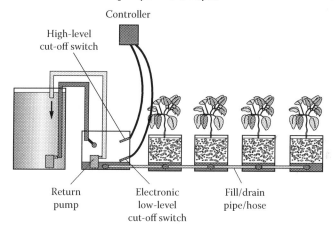

FIGURE 20.4 Ebb and flow system filling and draining cycles. (Drawing courtesy of George Barile, Accurate Art, Inc., Holbrook, New York.)

An option to this ebb and flow system is to use a top feed drip irrigation system. The grow pots are set up similar to the ebb and flow system with a drain tube connecting each pot to the distribution tank. Two float valves or electronic switches, one at the lower limit and the other at a high limit will operate the submersible pump that conducts the solution back to the nutrient reservoir as it returns from the grow pots. The difference in this system is that the submersible pump in the main solution reservoir irrigates the grow pots directly by a drip irrigation system rather than pumping to the distribution tank.

The system may be set up as a closed system whereby the returning solution is collected and conducted back to the distribution reservoir where it is pumped back to the main reservoir. Otherwise, with an open system there would not be a return drain line to a distribution reservoir, the leachate would drain to waste through the greenhouse floor.

Ebb and flow systems are available that use no substrate making them more of a water culture system (see Current Culture H2O in Appendix). The principles of operation are very similar to the one just described, but without a medium, growing the plants in an 8″ net pot that is irrigated from below. Nonetheless, even with this system it would be better to use expanded clay in the net pots. In that way, the transplants are kept higher in the pots so that their crowns are above the upper water level at all times.

DRIP IRRIGATION SYSTEM

A drip irrigation system is the basis of most hydroponic cultures. The specific cultures including coco coir, expandable clay, peatlite, perlite, rice hulls, rockwool, sand, sawdust, and various mixtures of these all use drip irrigation. The various cultures may be named by the substrate used, but overall they are all drip irrigation systems. While most commercial drip irrigation systems use an injector/proportioner component, in backyard greenhouses it is more economical to use a nutrient tank of normal strength solution than stock tanks of concentrated solution that is diluted by the injector. The drip system may be an open to waste or a re-circulation design. I recommend using a closed (re-circulation) system to prevent the solution from draining directly below the greenhouse, which may cause a moisture build-up, especially in northern climates during the winter months. When you change the nutrient solution it may be pumped to your garden and landscape plants, especially when those plants are actively growing.

The first component is a nutrient reservoir. In a backyard greenhouse with the peak being relatively low at about 9 ft, it is better to keep the plant growing systems at floor level. Locate the nutrient reservoir at the end of the greenhouse opposite the entrance door. Set the reservoir into the ground so that the top is 2″ above the floor of the greenhouse (Figure 20.5). Purchase a plastic tank of approximately 200–250 gal in volume, but it must not be a tall tank to facilitate burying it. Additionally, if the greenhouse is located in an area of high groundwater table, it could cause the tank to collapse. A more ideal tank is a rectangular one of about 18–20″ wide by 2 ft deep by 8–9 ft long. If you cannot locate such a tank, use a number of plastic storage tanks and join them by bulk-head fittings at the bottom of the sides. In that way, the modular tanks will in effect be one. Place 2″ thick Styrofoam on the bottom and around the sides of the excavated area between the soil and the tank to insulate it from the cold soil during winter months. Construct a cover of ¾″-thick plywood. Seal it with oil-based paint.

The rest of the system includes a pump, time-clock controller, piping, and the drip lines as shown in the diagrams (Figures 19.28 and 20.5). Use a submersible fountain pump or a sump pump operated by a controller. The volume (gpm) of the pump is calculated by the number of drip lines, which is a function of the number of plants. In our example hobby greenhouse, if we grew all of the plants in slabs or bato buckets the maximum number of vine crops the greenhouse could contain would be three beds with 12 plants each for a total of 36 plants. If each plant receives one drip line on an emitter of 0.5 gal per hour (gph), the total flow would be $36 \times 0.5 = 18$ gal per

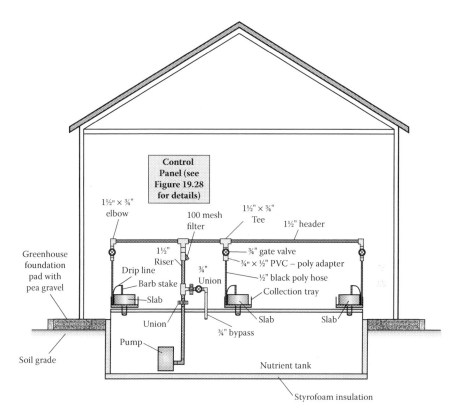

FIGURE 20.5 Cross section of nutrient tank at one end of greenhouse with components. (Drawing courtesy of George Barile, Accurate Art, Inc., Holbrook, New York.)

hour. Select a pump that has a flow capacity of twice that for lots of flexibility in crops, which would be about 40 gph. The next step in the pump selection is the lift it must satisfy. This is done by calculating the total frictional loss within the piping. These calculations are not presented as for such a small greenhouse a lift of 10–15 ft would be more than adequate. Finally, the pressure should be about 30 lbs per square inch (psi) to enable the operation of solenoid valves if you should wish to irrigate some of the rows with different cycles and periods of watering. This applies to the growing of numerous crops having different water demands. If this flexibility is preferred, a controller having 4–5 stations would have to replace the simple time-clock. I do not recommend this type of sophistication for a small backyard greenhouse.

To select the best value of pump with a low flow rate of 40 gph, the lift is inadequate due to its very small size. I prefer to select a larger pump and install a bypass line back to the solution reservoir to regulate the flow volume. There are small fountain pumps of 800 gph capacity with a lift of 12 ft for less than $50. These types of pumps would not produce sufficient pressure to operate solenoid valves. There are a number of "Little Giant" submersible pumps with a capacity of 70 gph up to 12 ft of lift with a pressure of 5 psi for about $100. They are 1/40 HP at 100 watts and draw 1.7 amps of electricity. Other models for about $150 have a capacity of 200 gph at 14 ft with a pressure of 7.5 psi with a ½″ male outlet. They are 1/15 HP at 200 watts and 3.2 amps. For pressure compensating emitters of 2 liters per hour (0.5 gph), a pressure of 14.5 psi is optimal pressure for their operation. However, at 7.5 psi the volume of water delivered is a little less, about 1.75 liters/h or 0.46 gph, so this is an acceptable rate of flow. At lower pressure a larger emitter could be used. For example, a 4 liter/h emitter would give you 3.5 liter/h (0.9 gph) at 7.5 psi or about 3 liter/h (0.8 gph) at 5 psi. Alternatively, the length of any irrigation cycle can be increased when using a lower volume delivery of water by the emitters under lower pressure. In general, it is best to be oversized in the pump and use the bypass to cut back on the flow volume. The maximum pressure for the drip emitters is two bars or 29 psi. At that pressure they will deliver twice the volume that they are rated at as the rating is based upon 1 bar or 14.5 psi of pressure. There are also larger submersible pumps such as the Little Giant one-sixth HP utility pump capable of 840 gph at 15 ft of lift. It has a 1″ discharge. It draws 5 amps and uses 380 watts of power. It costs about $150. This size of pump would be a better choice over the smaller models.

Some suitable pedestal sump pumps are available. The advantage of the pedestal sump pump over the submersible types of pumps is that the pump motor is not in the presence of the nutrient solution. Many submersible pumps eventually start to leak and break down due to the corrosive action of the nutrient solution. All impellers of any of these pumps must be either plastic or of stainless steel to resist corrosion. A 1/3 HP pedestal sump pump will deliver up to 2100 gph at 15 ft or 900 gph at 20 ft of lift with 8.65 psi. The electrical demands are 330 watts at just under 3 amps and require a 15 amp circuit. They have a glass reinforced nylon impeller.

Pedestal sump pumps cost between $80 and $100. These types of pumps are available at a building supplier, plumbing store, or online.

The next step is the piping from the pump to a header to which the drip irrigation lines are connected. The connection to the pump depends upon the type of pump and its outflow diameter. The submersible pumps mentioned earlier have ½″ or 1″ outlets. The pedestal pump has a 1¼″ outlet. The ½″ outlet needs a female adapter connection while the others use a male adapter fitting. If using the larger submersible or pedestal pumps install a bypass (Figure 20.5) to regulate the volume flow of the pump as it exceeds that needed for the drip system. Use flexible black polyethylene piping from the pump to a header pipe attached to the back wall of the greenhouse about 30″ above the top cover of the solution reservoir as shown in the diagram. Install a ¾″ diameter bypass line from the pump riser with a union and a gate valve in this line as it returns to the nutrient reservoir. Keep the gate or ball valve near the upper end of the pipe so that it is easy to adjust. Then, with an elbow, make the return to the tank. Immediately before connecting the inlet from the pump to the header install a 100 mesh filter. The header pipe should be 1½″ diameter schedule 40 PVC. Attach the inlet from the pump with a PVC tee. Assuming there are three beds of plants assemble the header with two 1½″ elbows (one on each end) and one 1½″ tee for the middle bed. Install a ¾″ ball valve downstream from the tee or elbows within 2″ of the fittings. The ball valves will allow the balance of flow to the drip lines. After the ball valves convert the piping to black poly from PVC using a slip thread reduced bushing converting from ¾″ to ½″ to adapt to the ½″ black poly drip hose.

Place the drip lateral line hose on top of the bato buckets or at the sides of the slabs. Use a 3″, 1″ diameter piece of PVC pipe to plug the end of the poly hose by bending about 6″ of it back and slipping the PVC piece of pipe over it as a sleeve. Alternatively, you may purchase a Figure "8" end stopper. Make up 18″ long drip lines from 0.160″ to 0.220″ diameter drip line. Punch the holes for the emitters in the poly hose using a special punch tool available from irrigation stores or online. The position of the holes can be on the top of the black poly hose for easy access. Locate them where the plants will be set. Insert the emitters and attach a drip line to each. Insert the other end of the drip line into a barbed stake that will keep the water from spraying on the plants. One stake will be placed on top of the rockwool block in which the transplant is growing.

If plant towers are located in one of the rows bring the ½″ black poly hose from the header up to the top of the plant towers and thread it through a 1″ tee at the top of each plant tower. The tee is fixed to the outer sleeve support of the plant tower as shown earlier in Chapter 13 (Figures 13.26 and 13.27). From the black poly hose insert three compensating pressure emitters of 4–8 liters/h (1–1.5 gph) above each plant tower. Make two drip lines long enough with sufficient slack to enter the top pot of the plant tower and the other one to reach the center pot of the plant tower. Insert a barbed stake at the end of each drip line to secure it into the substrate of the pot.

The plant towers are set up exactly as explained in Chapter 13 with the exception of the supporting frame dimensions. The plant towers in the greenhouse could drain to waste or the solution could be re-cycled by raising them up on a frame to an elevation above the nutrient tank (about 6″). The collection pot would conduct the solution to a return pipe going back to the nutrient cistern. In this case, construct the supporting frame to a height of 6″ or slightly higher to enable the spent solution to return to the cistern.

There are many systems to choose from, so select the ones that best suit the growing requirements of your plants. You can easily assemble a number of different hydroponic systems within a backyard greenhouse. Construct all of these systems yourself from components readily available from hydroponic shops and irrigation suppliers. Websites such as that of Grainger, Inc. offer their online catalog that contains heaters, fans, pumps, time-clocks, controllers, and many other components (see Appendix). Enjoy the fun of building your own systems and experimenting with different ones to determine which crops and systems work best together!

21 Sources of Supplies and Components for Backyard Greenhouses

Often supplies and components for backyard greenhouses may be sourced from local building suppliers, distributors for plumbing, electrical, and heating components. For specific greenhouse-related supplies and components contact greenhouse builders and distributors as well as hydroponic outlets. There are specialized manufacturers and distributors of hobby, backyard greenhouses, their components, and supplies. If you are in the market for a backyard greenhouse visit personally some of these companies as they will be able to assist in your decision as to the size, nature, and type of greenhouse most suitable for your needs. They can also recommend the components of heating, cooling, lighting, and so on to provide optimum growing conditions in a specific location. They can calculate the heating and cooling capacities for your area and recommend the size and type of these units to purchase. Consider capital and operational costs of various options of these components. The optimum levels of environmental factors are also a function of the crops grown. Normally, the capacities of the components are based on the extreme levels of the weather conditions. Under such extremes, the conditions within the greenhouse can be maintained at less than optimal, but adequate to prevent any damage to the plants. Extremes generally occur over relatively short periods during the early hours of the morning so will not have long-term effects on the health of the plants.

TEMPERATURE

Control of optimum temperatures is based upon heating and cooling. These two factors are regulated by heaters, ventilation fans, evaporative coolers, and shading (Figure 21.1). These components are available from many suppliers of backyard greenhouses (see Appendix).

HEATING

Smaller greenhouses up to 20 ft in length by 10–12 ft in width may be heated with space heaters. Longer greenhouses should use unit heaters with a convection tube that distributes the heated air rapidly down the length of the greenhouse. The unit heater has a fan at the back that blows the air over heat exchangers into the convection tube. This type of heating system is mounted in the peak area of the greenhouse and circulates heat above the plants. It is not as efficient as using hot water from a boiler through heating pipes at

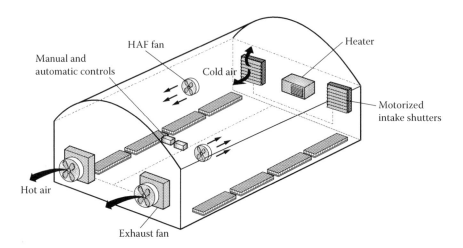

FIGURE 21.1 Temperature control in a greenhouse. (Drawing courtesy of George Barile, Accurate Art, Inc., Holbrook, New York.)

the base of the plants. But, such a hot-water heating system is very costly although much more efficient in providing more uniform heat throughout the crop and is less expensive to operate. A mixing fan or horizontal air flow (HAF) fan mounted just above the crop assists in mixing the air within the greenhouse. The heating systems discussed here are space heaters as most backyard greenhouses are of 10 ft × 12 ft or 10 ft × 16 ft and such a heating system is efficient and economical for the greenhouse size.

In Chapter 19, heating calculations and some websites that provide such information were presented. Projected costs of heating were also given. Emphasis here is on the types of heaters and their advantages or disadvantages along with cost ranges for the various types and models (Figure 21.2). Space heaters may be electrical or fired by propane or natural gas. Natural gas from your house would provide the most economical operational source of heat, but, the cost of installation is more expensive than to use propane or electrical sources. An underground cable must connect the greenhouse with the residence, so size it sufficiently to meet the demands of the space heater (usually 220 volts, 30 amps). A 220-volt heater is more efficient than a 110-volt one and at the same time has more heating capacity. Most of these heaters come with a two-speed fan. The 220-volt heaters, compared to 110-volt ones, have a stronger fan that can better mix the air in the greenhouse.

An alternative is to locate electric baseboard heaters along the sides of the greenhouse. The air will rise up and circulate as it cools. The disadvantage with this type of heat is that it is located very close to the outer rows of plants and could cause burning of leaves of those plants. The heat is focused in one area of the crop and can also reduce relative humidity to low levels near the site of the heaters. If baseboard heaters are installed, a HAF fan should be mounted near the top of the crop above one of the aisles or the center row to mix the air to create convection currents.

Infrared heaters of relatively small sizes are available for application in backyard greenhouses. The advantages of infrared heaters are lower electrical consumption, no noise from a fan, and they heat the plants and you, not the air. There are a number

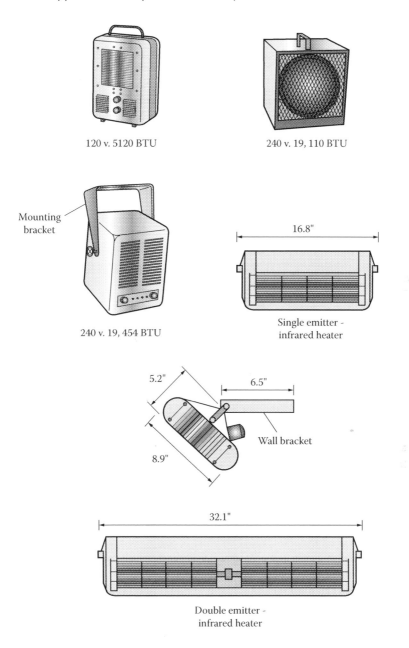

120 v. 5120 BTU

240 v. 19, 110 BTU

Mounting
bracket

240 v. 19, 454 BTU

16.8"

Single emitter -
infrared heater

5.2"

6.5"

8.9"

Wall bracket

32.1"

Double emitter -
infrared heater

FIGURE 21.2   Various heaters for backyard greenhouses. (Drawing courtesy of George Barile, Accurate Art, Inc., Holbrook, New York.)

of types, but all are mounted above the crop to maximize coverage. Some are natural gas or propane operated while others are electrical.

Modine manufactures several natural gas or propane small models of 30,000 and 50,000 BTUH, respectively. In our greenhouse example of 10 ft × 12 ft (see

Chapter 19), for weather conditions of Seattle, WA, the heating system capacity was 9700 BTUH. The Modine heaters would be too large for this size of greenhouse. Other types of infrared heaters are available as a modular design with single, double, or triple emitters (heater units) in one unit. These are electric and a selection of wattage, voltage, and amperage per unit gives versatility in satisfying numerous heating demands. The smallest one emitter is 1.5 KW and 12.5 amps going up to three emitters at 6 KW and 25 amps (240 volt). They are wall-mounted with special brackets or supported overhead with chains. These heaters may be mounted singly or in banks up to 3 units. The smallest unit (one emitter) mounted at 8 ft will cover a surface area of 10 ft × 10 ft × 12 ft to supply heat over 110 sq ft of surface. The two emitter model, mounted at 10 ft, covers 160 sq ft of surface heat zone. Prices range from $400 for a single emitter to $600 for 4 KW, 19 amp triple-emitter heaters. They are considerably higher in price than standard electric space heaters, but over time the savings in heating costs would compensate for the higher initial capital cost.

The smallest 120-volt electric portable greenhouse heater capable of generating 5120 BTU can heat up to 120 sq ft of greenhouse. It comes with a thermal cut-out to prevent overheating. However, a single-stage thermostat could be wired in line to operate the heater. This size of heater is too small for a 10 ft × 12 ft greenhouse. I have found it is good for an 8 ft × 12 ft free-standing or lean-to greenhouse. This type of heater is available for just under $50. A larger 240-volt, industrial electric heater produces up to 19,000 BTU. It is 5600 watts, 23 amps with a built-in thermostat, and thermal safety limit switch to prevent overheating. This size of heater easily fulfils the heating needs of our 10 ft × 12 ft greenhouse example and would be adequate for a 10 ft × 16 ft greenhouse. It costs about $150.

Natural gas or propane greenhouse heaters will generate from 20,000 to 25,000 BTU. These are suitable for larger backyard greenhouses of at least 10 ft wide by 20 ft long. However, they are also recommended for greenhouses of 150–175 sq ft of floor area in more northerly locations. They are available as vented or non-vented models. If using a nonvented heater the source of natural gas or propane must be clean burning not to create any carbon monoxide or other byproducts that could damage the plants. For this reason, it is safer to use a vented model. All gas-burning models must have a fresh air intake to provide oxygen for complete combustion. A 2–3″ PVC intake pipe mounted within 1 ft of the floor at one side of the heater is recommended. Place an elbow oriented downward on the outside with a screen cover to prevent water and rodents from entering the greenhouse. Prices, based on size, vary from $450 to $600.

Ventilation

The next step in temperature control is ventilation to assist in cooling the greenhouse. Ventilation is achieved through the use of exhaust fans and circulation fans. The size of ventilation fan is determined by the volume of air in the greenhouse to be exchanged once per minute, which is expressed as cubic ft/min (CFM). CFM is expressed as: CFM = Length × Width × 12. Our example greenhouse has a needed air volume exchange of: 12 × 10 × 12 = 1440 CFM. Select an exhaust fan based upon that air exchange rate. Ventilation packages are available that include a shuttered exhaust fan, air intake shutters, and a thermostat. A 16″ shutter fan, two 18″ intake shutters and a thermostat costing about $460, has a 2000 CFM rating. It is best to install a variable speed fan that operates from a two-stage thermostat or a variable speed controller that initiates different speeds of the fan. The advantage to this type of fan is to enable minimum cooling during sunny days with ambient temperatures below optimum for the greenhouse crops. Due to the "greenhouse effect," the greenhouse air temperature heats up fast under sunny conditions regardless of the outside temperature. The exhaust fan starts at a lower speed to exchange the air slowly, but keeping it within optimum levels. If the temperature in the greenhouse continues to rise above optimum, a faster fan speed will be initiated. This prevents a rapid influx of cold air into the greenhouse. The slowly incoming air will mix with the inside air keeping it optimum around the plant canopy.

Locate the exhaust fan on the north end of the greenhouse above the nutrient tank at about 7 ft from the ground. Installation instructions are given in Figure 21.3. Mount the fan frame on horizontal support bars bolted to the vertical framework of the greenhouse. The motor on the fan is on the inside of the greenhouse with the shutters outside. Caulk the perimeter of the shutter to the greenhouse glazing. The two intake shutters are situated on the opposite end, one on each side of the door, about 18″ above ground level. These intake shutters may operate with the pressure difference created during the exhaust fan activation as shown in Figure 19.23 or they may be regulated by a small motor that is coordinated with the operation of the exhaust fan.

Thermostat and speed control units should be mounted at plant height near the center of the greenhouse so as not to be influenced by drafts directly in line with the exhaust fan or heater. Protect them from direct sunlight by mounting them to a

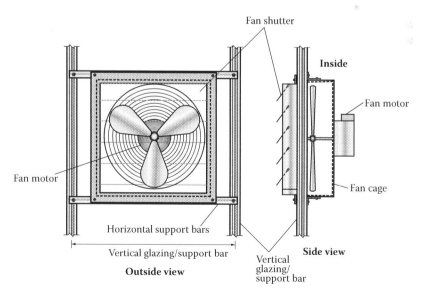

FIGURE 21.3 Installation of exhaust fan in greenhouse end wall framing. (Drawing courtesy of George Barile, Accurate Art, Inc., Holbrook, New York.)

piece of plywood painted white. Put a small top above them to shade them from the sun as shown in Figure 21.4.

Support one small circulation fan (10″ diameter) above each aisle between the crop rows. Locate one next to the door facing the opposite direction and the other at the exhaust fan end in front of the nutrient tank (Figure 21.5). These fans will circulate and mix the air to make the temperature more uniform throughout the greenhouse. They cost less than $100.

As mentioned in Chapter 19, many backyard greenhouses have solar operated roof vents. This is a less expensive method of ventilation than forced air that depends

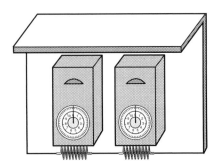

FIGURE 21.4   Thermostats with top cover. (Drawing courtesy of George Barile, Accurate Art, Inc., Holbrook, New York.)

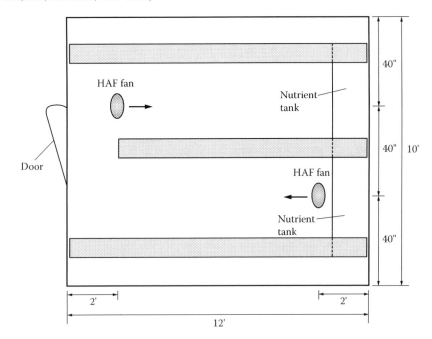

FIGURE 21.5   Location of horizontal air flow (HAF) fans. (Drawing courtesy of George Barile, Accurate Art, Inc., Holbrook, New York.)

entirely upon convection air currents within the greenhouse. Under high summer sunlight and temperatures, natural ventilation may be inadequate. Under such conditions, cooling can be assisted greatly by the use of an evaporative cooler or the use of an evaporative cooling pad on the end of the greenhouse opposite the exhaust fan.

Cooling

There are two methods of cooling: evaporative cooling and fogging systems. With an evaporative cooling pad two exhaust fans are part of the system. Locate the evaporative cooling pad on the north end above the nutrient tank and two exhaust fans on the opposite, entrance end of the greenhouse. If the greenhouse has natural solar vents, they must be closed (disabled) during a cooling cycle to prevent short-circuiting of the air entering directly from the roof vents and bypassing the cooling pads.

The size of the cooling pads is a function of the total CFM of air to be exchanged. In our 10 ft × 12 ft greenhouse, we needed 1440 CFM. For 4″ thick cooling pads use 1 sq ft of pad per 250 CFM. That would be: 1440/250 = 6 sq ft. The smallest self-contained cooling pad systems come 3 ft wide by 5 ft long and are capable of cooling about 4000 cubic ft. So, such a unit would be more than adequate for our 120 sq ft greenhouse.

However, smaller units can be made upon request from suppliers. The reservoir, collection pipe and inflow pipe of a self-contained unit are constructed of polyvinyl chloride (PVC). This eliminates the need for a separate external sump tank as it along with all the piping (outflow distribution pipe over the top of the pad and the bottom return pipe) connect to a small PVC reservoir that has a submersible pump as shown in Figure 21.6. This could be constructed as a do-it-yourself (DIY) project by simply purchasing some 4″ thick cooling pads of 2-ft or 3-ft tall and make it 4–5 ft long. The cooling pad panels are 12″ wide by 2–8 ft tall in 1-ft increments. It would be best to use either 2 ft or 3 ft tall sizes. Locate the cooling pad in the middle of the north end of the greenhouse or that end above the nutrient tank. Place it to one side of the control panel 3–4 ft above the floor. A ready-made cooling pad system of 3 ft × 5 ft costs about $800.

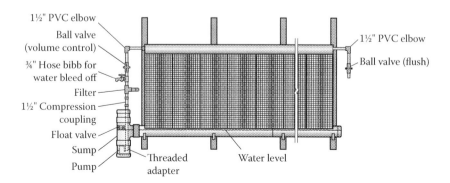

FIGURE 21.6   Small pre-assembled cooling pad for backyard greenhouse. (Drawing courtesy of George Barile, Accurate Art, Inc., Holbrook, New York.)

It is best to choose two exhaust fans that have a larger capacity than the 1440 CFM as the static pressure though the cooling pad reduces the air flow slightly. Therefore, the total rate of air exchange of the two exhaust fans should be about 2500 CFM.

An alternative to using a cooling pad system is to use an evaporative cooler (Figure 21.7). This is the preferable method for small greenhouses. They come in numerous sizes ranging from 1000 to 6500 CFM with prices ranging from $400 to $1300. Two types available include through wall and ducted models. The coolers are situated outside the greenhouse at the north or tank end wall. It is best to support the cooler on a concrete pad to keep it free of dirt and debris. Provide access through the greenhouse covering to fit the discharge of the cooler. This cooler is self-contained with a blower that moves outside air across cooling pads inside the unit and pushes cold air into the greenhouse. As a positive pressure system, it forces the hot air in the greenhouse out roof vents or exhaust shutters as shown in Chapter 19 (Figures 19.25 and 19.26). The system is operated by a two-stage thermostat that runs the cooler with or without wetting the pads. The first stage is a dry pad and as the temperature reaches a second limit, the pad is moistened to further reduce the temperature. The coolers come with a water hose connection for constant water supply. Be sure to match the CFM ratings of the cooler with the total of the exhaust shutters or roof vents. Be careful to winterize the evaporative cooler by draining all water from the system before any threat of frost.

A very simple method for additional cooling is to attach a mist ring on the outside protector cage of circulation fans within the greenhouse (Figures 21.8 and 21.9). These mist rings emit a fine mist (1/2 gph per nozzle) that is dispersed throughout the greenhouse by the fan cooling the air while also adding humidity. Attach the mist ring to a standard water hose. They are available in three sizes depending upon the fan size, using 3, 4, and 5 nozzles per ring. The ring is easily attached to the fan cage by use of plastic zip ties. Connect a water filter on the garden hose or hose bibb in line with the mist ring to prevent clogging of the mist nozzles with debris. The mist rings range in price from $25 to $30 for the different sizes.

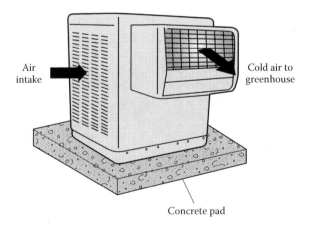

FIGURE 21.7 Evaporative cooler mounted outside of greenhouse. (Drawing courtesy of George Barile, Accurate Art, Inc., Holbrook, New York.)

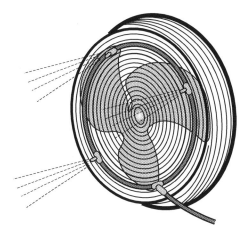

FIGURE 21.8   Evaporative cooling by mist ring on a circulation fan. (Drawing courtesy of George Barile, Accurate Art, Inc., Holbrook, New York.)

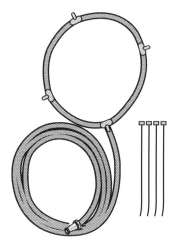

FIGURE 21.9   Mist ring for a circulation fan. (Courtesy of CuisinArt Golf Resort and Spa, Anguilla.)

Fogging is an efficient cooling method, but unfortunately most systems are expensive and more applicable to large greenhouses. Such systems operate at a water pressure of 1000 psi in order to produce droplets of 5–45 microns. At that size, the water particles remain suspended in the air as a fog, not a mist. The evaporation into the air of the water lowers the air temperature. To generate such fine water droplets a high pressure pump, flexible tubes, filter, and nozzles make up the components of the system. Even the smallest portable units are for an air volume of 5000 cubic ft, about four times that needed for our 10 ft × 12 ft example backyard greenhouse. The fogging nozzles are attached to the cage of a very high velocity fan that would damage the crop in a small greenhouse. Other small portable humidifiers are those used in homes to increase the relative humidity in the air. These can be used in a backyard

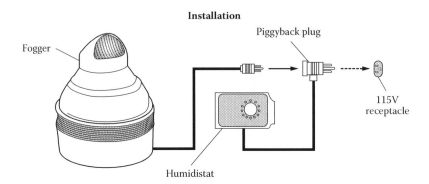

FIGURE 21.10 "Hydrofogger" humidifier for cooling a backyard greenhouse. (Courtesy of Hydrofogger.com, P.O. Box 31281, Greenville, SC 29608.)

greenhouse. They are operated by a humidistat located 5–6 ft above the floor. The humidistat is connected in series with the fogger. It will control the fogger within a 20%–80% humidity range. A typical humidifier of this type is the "Hydrofogger" at www.hydrofogger.com (Figure 21.10). This unit is recommended for small greenhouses and indoor gardening with air volumes of up to 4500 cubic ft. They also have a smaller "Minifogger," which covers up to approximately 1080 cubic ft and a "Cuoghi Mini NEB" that covers up to 1980 cubic ft. Prices range from $300 to over $600.

SHADING

Partial shading, up to 40%, is recommended during the longest, hottest days from late spring to early fall. The shade cloth can be placed on top of the greenhouse roof. The shade will assist in reducing some of the heat entering the greenhouse and therefore save on the evaporative cooling. Avoid resting the shade cloth directly on the greenhouse covering, especially if the glazing is glass or polycarbonate. Place a few bars across the roof to keep the shade several inches above. This will permit some ventilation between the shade and the glazing and will avoid any possible adhering of the shade cloth to the glazing under high temperatures. If you do not want to cover the greenhouse with shade cloth, use whitewash paint over the hottest months. It can be washed off easily in mid-fall so as not to limit light as the days shorten. An excellent product is "ReduSol" by Mardenkro (www.mardenkro.com).

LIGHTING

Supplementary artificial lighting is essential to increase yields and shorten cropping cycles during the short day lengths and low light levels of late fall through winter to early spring months. Plant growing lights include cool white, high-output fluorescent, high intensity discharge (HID), light emitting diode (LED), and compact fluorescent lights as shown in Figures 21.11 through 21.14. The choice of light is a function of light demand of the specific crops, capital cost, operational cost, and the area served by a given unit.

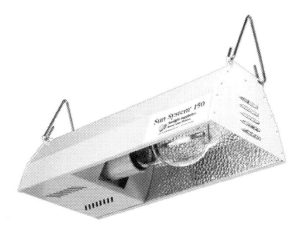

FIGURE 21.11 High intensity discharge (HID) light and fixture. (Courtesy of Sunlight Supply, Inc., Vancouver, Washington.)

FIGURE 21.12 T5 fluorescent lights. (Courtesy of Hydrofarm Horticultural Products, Petaluma, California.)

There are two types of high intensity discharge (HID) lights: high-pressure sodium (HPS) and metal halide (MH). The HPS lights provide more energy in the red part of the spectrum, which promotes flowering and fruiting, while the MH lights are more intense in the blue causing rapid growth. HPS lights are better to

FIGURE 21.13 Light emitting diode light. (Courtesy of LumiGrow, Novato, California.)

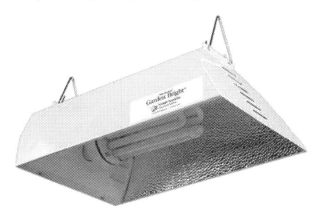

FIGURE 21.14 Compact fluorescent light and bulb. (Courtesy of Sunlight Supply, Inc., Vancouver, Washington.)

supplement sunlight to extend the day length or to increase intensity during cloudy periods whereas MH light is more useful for indoors without natural sunlight. These lights give off considerable heat so must be mounted about 3 ft above the crop. With the relatively low eave height of backyard greenhouses, it is not possible to mount these lights above the crop. The alternative is to mount one unit near the peak of the roof in the middle of the greenhouse. This position of the light unit reduces the efficiency of the light distribution as the plants directly below will receive more light than the rows next to the sides of the greenhouse. Another disadvantage is that the lights are expensive to purchase and operate.

For small backyard greenhouses fluorescent lights are more efficient. The new T5 lights put out more intensity than older fluorescent lights. They have the advantage of offering higher light efficiency with low heat and provide a full spectrum in the red and blue for plant growth, flowering, and fruiting. They are available as a high output tube that provides about twice the intensity of light than the standard T5 tube. They are available as single tubes or multiple tubes, up to 8, in a 4-ft reflector unit (Figure 21.12). A single unit fixture and bulb costs about $50. Although they do not have the intensity of HID lights, they can be placed within 6–18″ of the plant due to cooler output and therefore can give similar intensity at plant leaf surface. These fluorescent lights are lighter in weight and have lower profile. Their ballasts may be remotely located to further reduce any heat generated directly above the plants.

LED lights are very light in weight, generate very little heat, and have very long life. They maximize red and blue light so are well balanced for the vegetative and flowering growth of plants. However, to get sufficient intensity and uniform distribution of the light it is necessary to purchase LED arrays where many LED lights of white, red, and blue combinations fit within a reflector unit (Figure 21.13) that will produce an equivalent of a 1000 watt HID lamp. Prices of these arrays are high from $1000 to $1200 each.

How much lighting is needed? Most vegetable crops need 50–70 watts of light per square foot of growing area if the artificial light is the only source of light. In a backyard greenhouse that has natural light also available, the level of supplementary lighting can be reduced to one-quarter of that. Calculate the correct wattage of light needed for a specific area by multiplying the desired wattage of the light by the area in square feet. In our greenhouse example of 10 ft × 12 ft the total area is 120 sq ft. So, we need at least 50 watts × 120 sq ft × 25% = 1500 watts. One 1000-watt HID light mounted near the peak of the greenhouse would provide this amount of light as would an LED light array. My preference for a backyard greenhouse is to use three 8-ft, dual tube, high-output fluorescent units, one above each row of plants. With the 4-ft T5 high output four tube fixtures use a total of 6 units. The cost for those would be about $800.

Support the fluorescent lights about 18″ above the tops of the plants using jack chains that will allow their adjustment as the crop grows.

Compact fluorescent lights with fixtures (Figure 21.14) (one bulb per fixture of 125 watt) will cover a maximum of 3 ft × 3 ft of growing area. For our greenhouse example we would need 4 units per row or a total of 12 units. These cost about $70 a unit, so while the lights last about 10,000 h the number of units needed is not practical compared to other light sources as the cost for 12 units is over $800.

With a single unit of HID or LED array, it is advantageous to use a light mover that moves the light back and forth on a track above the crop to give better light distribution over time (Figures 16.5 and 16.6). This gives better light penetration throughout the crop than just those plants immediately below the light unit. Supplementary light is also for extending the day length during the winter months.

Keep day length at 12–14 hours, as plants need a period of darkness. In addition, you do not want to upset any neighbors by lighting the greenhouse beyond normal summer months of daylight. Operate the lights with a time-clock, starting them about 8:00 AM and turning them off by 10:00 PM. That will give 14 h of supplementary light. On days with sunlight, turn off the lights by 9:00 AM and activate them again about 3:30 PM in the afternoon as the sun intensity falls. It is also possible to get a photo sensor that could regulate the operation of the lights.

CONTROLLER/TIME-CLOCK FOR IRRIGATION

While the exhaust fan, roof vents, and heaters can be controlled with thermostats, the irrigation cycles are operated by a time-clock or controller. Irrigation controllers offer more functions than a simple time-clock, but are relatively expensive. Time-clocks should be of 24-h periods with intervals capable of separate hours and minutes. There are electronic 7-day programmable timers having 1 min on/off cycles with up to eight cycles per day for about $25 (Figure 21.15). This programmable timer is ideal for activating irrigation cycles. For about $15 you can purchase a mechanical timer that has 15 min on/off cycles with a 24-h time period (Figure 21.16). This type of timer is fine for turning lights off and on, but is not suitable for irrigation cycles. A time-clock will operate as only one station, so it does not have the capability to operate different irrigation cycles and periods for different rows of crops. To achieve this use an irrigation controller having at least three to four stations. To operate independently several stations solenoid valves are installed within each irrigation line to the plants. These would be located near the header pipe after the ball or gate valve,

FIGURE 21.15   Electronic programmable time-clock. (Courtesy of Hydrofarm Horticultural Products, Petaluma, California.)

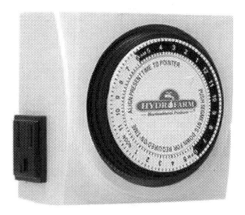

FIGURE 21.16   24-h, 15-min increment mechanical timer. (Courtesy of Hydrofarm Horticultural Products, Petaluma, California.)

but before the conversion to the black poly hose having the drip lines. These solenoid valves would be for ¾″ diameter pipe as that is the pipe size from the header.

Many irrigation equipment suppliers, such as Hunter Industries, Rain Bird, and Toro have controllers for home landscapes (See Appendix). The least expensive controllers (under $300), which have four stations, have only four starts per station during a day. That is not enough for irrigating during hot days when you may need at least 8–10 starts.

There are also timer switches (time-clocks) at lower prices from $25 to over $250 that turn one output on or off over a period of 7 days and can be adjusted to within 1 min intervals. The electronic versions are the expensive ones. Electromechanically operated timers are at the low price range. Intermatic (www.intermatic.com), Tork, and Paragon make both mechanical and electronic timers that vary in price from $100 to over $350. They are available through Grainger, Inc. (www.grainger.com) as are many other components as fans, heaters, vents, and so on. The lowest price time-clocks are 7-day clocks that can activate a circuit for a minimum of 15 min.

Using a 2-pole relay, these timers can activate one 110–120 volt circuit (pump) and a 24-volt circuit (solenoid) at the same time using a step down transformer in parallel with the solenoid, but they act as only one station. This restricts their ability to operate the pump and two solenoids at different times as that would require a two-station controller or two independent time-clocks. Each electronic time-clock costs about $175. The low-priced timers are used to turn lights on and off, such as those used in households that are available for $25 or less.

After examining many of these controllers and timers, I believe that a low-priced timer is fine for operating the lights but cannot be used for irrigation. Overall, the best procedure is to operate all of the plant rows at the same time on one time-clock that activates the pump only. When growing all vine crops with drip irrigation on a small scale, there is no significant benefit of increased yields by using different irrigation cycles for different plant rows. Grow low-profile plants such as herbs, lettuce, and basil in a separate nutrient film technique (NFT) or raft culture system that is operated independently of the vine crops. If you wish to grow these crops in plant towers with a drip irrigation system, keep them on the same cycles as the vine crops operated by the same pump. The 7-day programmable timer for about $25 is suitable for these needs.

While some of these components seem to be complicated in their specifications for your specific needs, salespersons in companies distributing the products can assist greatly in simplifying your choices to suit your greenhouse needs. These are all on-the-shelf products available at many distributors in stores or online.

Section VI

Vegetable Crops and Their Cultural Techniques

22 Most Suitable Crops and Varieties

While we can grow literally all crops hydroponically, our choice largely depends upon economics. Under research conditions, many crops are raised using hydroponic culture, but the purpose of the research is to determine other physiological, pathological, nutritional, and/or pest effects on the plants under controlled atmospheres. The emphasis is not on hydroponic techniques, but only to use it as a tool in the research undertaken in other plant aspects. For such work, the economics of hydroponics is not at question for a particular crop. Crops such as cereals, rice, corn, beans, other legumes, root crops (carrots, potatoes, etc.), and many flowers and ornamentals have all been grown hydroponically during research studies.

In our case of growing on a small scale either indoors or in a backyard greenhouse, the choice of crop is not fully dependent upon economics, so noncommercial greenhouse crops may be cultured. For example, if you are very fond of some crops that you do not like the quality in supermarket products, grow them yourself hydroponically. Such crops as Asian greens, beans, beets, bok choy (Pac Choi), broccoli, Brussels sprouts, cabbage, carrots, cauliflower, celery, Chinese cabbage, kale, Kohlrabi, leeks, melons, okra, onions, onion sets, peas, radish, spinach, squash, Swiss chard, and zucchini will thrive in hydroponics. However, the choice of substrate is important. For instance, any root crops such as beets, carrots, onions, and radish will need a peatlite or sand substrate as that type of substrate permits the radical or bulb to form uniformly. Do not be afraid to experiment with various substrates and nonconventional crops. After all, part of the exercise is for it to be an enjoyable hobby. In this chapter, most of these nonconventional hydroponic crops are not discussed. Simply select such crops from a reputable seed company catalog. For backyard greenhouses seek varieties that will withstand high temperatures during summer months and others that do well under lower temperatures during winter months. Indoors most varieties will be suitable.

ARUGULA

There are two types of arugula: the roquette (rocket) and wild forms. The roquette is the standard salad arugula. The wild forms tend to be more pungent in flavor. "Astro" is a heat-tolerant variety of roquette form. "Sylvetta" also known as "wild rocket" is slower growing with a more pungent flavor. All grow well in all types of hydroponic culture.

BOK CHOY

This Chinese cabbage is also known as Pak Choi, Pac Choi, and Bok Choi. Most are green, but a few have purple leaves and red/green stems. Some varieties are tall, while others are more compact. My preference for growing hydroponically is the dwarf varieties as they will not lodge (fall over) or require large spacing and do well in plant towers (Figure 22.1), nutrient film technique (NFT), and raft culture systems. The smaller varieties can be spaced similar to lettuce (four plants/ sq ft). Some of the best varieties we have found to do well under hot conditions include: Green Fortune, Red Choi, and Takuchoy (Takuchoi). These are all dwarf in form. Other low-profile varieties (4–6″ tall) include the following: "F1 Hybrid Dwarf Bok Choy" that tolerates high heat and cold can be harvested within 45 days from sowing; "Dwarf Bok Choy" matures in 40 days; "Extra Dwarf Bok

FIGURE 22.1   Bok choy in plant towers. (Courtesy of CuisinArt Golf Resort and Spa, Anguilla.)

Choy" is only about 2″ tall and is ready in 30 days; and "Huo Guo Chai" is good for stir fry and soups. These varieties are available from Tainong Seeds, Inc. (www.tainongseeds.com).

LETTUCE

Iceberg lettuce should not be grown for several reasons. Firstly, it has very low nutritional value compared to bibb or leafy lettuces. Secondly, it does not form a tight head under high temperatures that normally would be encountered in a backyard greenhouse. There are many choices of varieties of leaf lettuce, oakleaf, Lollo Rossa, butterhead (Bibb), and Romaine/Cos. Leafy varieties include the green varieties "Black Seeded Simpson," "Domineer," "Malice," "Frizella," "Waldmann's Dark Green" and red varieties "Red Sails," and "New Red Fire." Red oakleaf lettuces are "Navarra," "Oscarde," "Red Salad Bowl," "Ferrari," and "Aruba." Some green oakleaf lettuces available are "Cocarde," "Tango," and "Green Salad Bowl." Lollo Rossa types comprise of "Dark Red Lollo Rossa," "Locarno," "Revolution," and "Soltero" to mention a few. Butterhead or bibb lettuces that I have grown with NFT and raft cultures include "Buttercrunch," "Charles," "Cortina," "Deci-Minor," "Milou," "Ostinata," "Rex," "Salina," and "Vegas." I have found that "Rex" is one of the best varieties for the raft and NFT systems under high temperatures as it bolts (goes to seed) slowly. Romaine or Cos varieties need cooler temperatures, and should hence be planted during the cooler seasons, otherwise they may bolt quickly under high temperatures. You may plant them under most temperatures if you do not require a crisp head formation. They will be more like a leafy lettuce under higher temperatures. Varieties include "Green Forest," "Parris Island," "Outredgeous" (red), and "Rouge D'Hiver" (red). Several fine leaf lettuces that are becoming popular are "Multi Red 1, 2, 3" and "Multi Green 1, 2, and 3."

There are numerous others to choose from presented in seed catalogs. The preceding are varieties that I have had success with growing hydroponically.

LETTUCE SALAD MIXES

These premixed blends of color and texture of lettuces are for mesclun mixes. The varieties in each mix are selected to have similar growth rates to allow uniform size at harvest. You may also create your own lettuce salad mixes. Simply, purchase the varieties you wish and mix them in a container before sowing them. You may mix different ratios of these varieties. For example, if emphasis is on red varieties use a larger portion of the seeds as red varieties.

Johnny's Selected Seeds (www.johnnyseeds.com) offers a number of pre-mixed lettuce blends. Their "Encore Lettuce Mix" has Green Oakleaf, Red Oakleaf, Green Romaine, Red Romaine, Lollo Rossa, Redleaf, and Bibb lettuces. The "Allstar Gourmet Lettuce Mix" combines Green Oakleaf, Red Oakleaf, Green Romaine, Red Romaine, Lollo Rossa, and Redleaf lettuces in a balance of color and texture proportions. The "Five Star Greenhouse Lettuce Mix" is a blend of downy mildew-resistant varieties for indoor culture. The "Wildfire Lettuce Mix" was created for high color contrast of the red varieties paired with green varieties.

These should be seeded in beds using a peatlite substrate. Shake the package before sowing to mix the seeds evenly. Sprinkle about 60 seeds per foot in a 2–4″ wide band. Cover lightly with about ⅛″ of peatlite or vermiculite medium. Keep the bed moistened with raw water until germination occurs and the seedlings reach about 1″ high. Thereafter, water every day with a half-strength nutrient solution using a watering can. Within 3 weeks, harvest by clipping the tops with a scissors. Do not cut lower than 1″ from the plant base to allow re-growth. Several, up to three, harvests may be made if cared for properly.

BASIL

There are many varieties of basils; the choice depends upon what you wish to use it for and what kind of flavor you want. The most common is the classic Italian sweet basil (Genovese). This variety is tall (24–30″) with large leaves up to 3″ long. It takes about 68 days to maturity, but may be harvested after 3 weeks from sowing. The seeds germinate in 5–10 days.

Basils need to be cut often, usually every few weeks, to keep the plants from flowering and going woody. The first cut should be made above the second node leaving at least two sets of side shoots below (Figures 22.2 and 22.3). That will train the plant to branch often so that it becomes very bushy. The next cut should be these side shoots again as they reach the second node. Harvest above the second node permitting four more side shoots to form. This constant pruning at this stage of side shoot development will keep the plants vegetative and reduce flowering. When flowers form immediately, pinch them to prevent the plant from getting generative and becoming woody. Eventually, after about 3 months, it is best to replace the plants. Start seedlings in a peatlite substrate or in Oasis or rockwool cubes depending upon the hydroponic system you are using. If in raft culture or NFT, it is best to use 1″ rockwool cubes. If in a bed of peatlite mix sow the basil in a peatlite mix in 72-celled compact trays. Sow seeds 3 weeks prior to the transplanting date at which you remove the old plants.

FIGURE 22.2   First cut of basil.

FIGURE 22.3   Basil after first cut.

It is best to use *Fusarium*-resistant varieties for indoor or greenhouse culture. Several Italian basils with disease resistance are "Aroma 2" and "Nufar." These seeds are available as "organic" as well as pelleted (clay coating) of inert National Organic Program compliant materials (Figure 22.4). Pelleted seeds are coated with a clay-base material that enables easier sowing and at the same time maintains moisture around the seed to prevent desiccation during germination.

Asian or Thai basils are used as a condiment in Thai and Vietnamese dishes. Several varieties are "cinnamon" and "sweet Thai" basils. These take about 64 days from seeding to maturity. Their leaves at 2″ are somewhat smaller than Italian basils. Often the blooms of these basils are used for flower bouquets. Cinnamon grows to 26–30″ tall while the sweet Thai is shorter at 12–18″. Care of the plants and the hydroponic systems are the same as for the Italian basils. There are citrus basils, "Lime" and "Mrs. Burns Lemon," that have a distinct citrus flavor and aroma. They are used to add a citrus flavor to fish and salads. The leaves are about 2″ long and the

FIGURE 22.4 Pelleted lettuce seeds.

plants grow from 20″ to 24″ tall taking 60 days to maturity. Once again grow and train them as for Italian basils.

Fine leaf or Greek basils are used like regular basil in pesto, soup, stuffing, or any vegetable dish, especially beans, peppers, eggplants, and tomatoes. Their flavor is stronger than Italian basils. "Spicy Bush" basil matures in 70 days with 1″ long leaves. Plants can be grown in pots or beds reaching a height of 8–14″. "Pistou" basil is a compact form of basil ideal for container growing. Leaves reach ½″ long with a height of 6–8″.

Purple basils may be used for garnishes and cut flowers. Some varieties are "Dark Opal," "Red Rubin," and "Purple Ruffles." These require about 80 days to maturity. Leaves are about 3″ long and the plant height reaches 16–24″. These purple basils normally contain up to 20% of green or variegated plants. I have grown them successfully in a peatlite bed as well as in NFT hydroponic systems. They will grow in plant towers, but the taller basils may lean over as they mature, causing shading of the plants below. As a result, it is not the best hydroponic system for basils apart from the low-profile varieties. Plant training and harvesting procedures are the same for all basils. Of course, if you want them to flower they must not be cut back as often as if they are to be used for the leaves and stems only.

HERBS

Most herbs grow well in peatlite beds and in plant towers using coco coir, mixes of peatlite with perlite or rice hulls, and in perlite by itself. Select more compact forms or those that hang down in the plant towers. Some of the most successful types that

I have grown in plant towers with a perlite substrate include chervil, regular chives, garlic chives, cilantro, dill, fennel, lavender, marjoram, mint, oregano, moss parsley, Italian (flat leaf) parsley, rosemary, sage, savory, and thyme.

"Vertissimo" chervil is slow-bolting and vigorous, and is hence good for greenhouse culture. It has a mild, sweet anise flavor and is popular for salads, micro greens, and garnishing. It takes about 2 months to first cutting. Chervil may be harvested for 2–3 months depending upon the weather. Under hot, summer conditions, it should be replaced after 3–4 months from sowing. The first harvest can be as early as when the plants reach about 4″ in height. Permit them to re-grow 4″ or taller before the next cutting. Chervil germinates in 10–14 days.

There are two types of chives: the "Fine Leaf" and the "Chinese Leeks/Garlic Chives." The chives have a mild onion flavor. The glove-shaped flowers may be used as an edible garnish. "Fine Leaf" chives have slender round leaves and are for fresh use. They take 75–85 days from sowing to first harvest. If properly maintained by frequent harvests, thereafter they will continue producing for almost a year. "Chinese Leeks/Garlic Chives" have thin flat leaves with a delicate garlic flavor. The flowers are edible and are used in bouquets. They take about 90 days to mature from sowing; however, you may begin cutting the leaves once they reach 6–8″ in length, but do not cut them back more than 3″ from the plant base (crown). This length allows sufficient photosynthetic leaf area for the plant to quickly re-grow. Chives germinate within 7–14 days.

Cilantro/coriander is very easy to grow in plant towers in bunches at the corners of the pots. All herbs that are directly sown in the plant towers should have 8–10 seeds sown at each corner to obtain a "bunch" of plants. Cilantro can also be grown as a micro green. The foliage of this plant is known as "cilantro," while the edible seed is "coriander." Some popular varieties are "Calypso" and "Santo" that grow to leaf harvest within 50–55 days from sowing reaching 12–18″ tall. These varieties are slow to bolt so are ideal for greenhouse hydroponic culture. They may be grown using plant towers or NFT. Cilantro seeds are actually fruits that contain two or more seeds. "Santo" seeds are available as "monogerms" whereby the seeds (fruit) have been split to get individual seeds permitting more precise planting. Also, the split seeds germinate faster than the regular fruit seeds. Sow six to eight seeds directly in the plant towers allowing 7–10 days for germination. If you cut the cilantro early when it reaches about 8–10″, two to three harvests will be possible before it goes to seed (bolts). Change the crop about every 3–4 months.

Dill may be grown for the foliage only or for the flowers. They take 40–55 days for first leaf harvest as shown in Figure 22.5, or 85–110 days for seed. Dill takes 7–21 days for germination. "Bouquet" grows to 38–42″ at maturity. However, to cut it for leaves in cooking begin harvesting when the plants reach 8″ tall. Cut it back lightly allowing about 4–5″ growth for regeneration. If you keep it cut back lightly each time it re-grows to 8–10″ tall, the plant will last for about 3–4 months. For use in making pickles, and so on, where the flower is needed, only one harvest is possible as the plant has matured at that time. "Fernleaf" is dwarf dill slow to bolt. It reaches 26–32″ in height. It has dark blue-green foliage and is best for growing in containers or in the plant towers. "Verling" dill is better for ornamental or cut-flower use. The leaves may be used for garnishing and culinary purposes. The plant grows to about 42–48″. Sow 10–12 seeds in the corners of the plant towers.

FIGURE 22.5   Dill in plant tower.

Fennel has feathery foliage with a sweet flavor used in salads, Cole slaw, and dressings. There are "Bronze" and "Bronze and Green" varieties. The seeds germinate within 7–14 days and mature within 50–60 days to a height of 24–36″. Sow four to six seeds per pot corner in the plant towers. These also should be harvested numerous times from when they reach about 8″ in height. Keeping them trimmed will permit three to four harvests between cropping over a 3–4 month period.

Lavender prefers a well-drained substrate and so does well in perlite in the plant towers. It will grow for a year between crop changes. Germination occurs within 14–21 days growing to a plant height of 12–30″. Sow four to six seeds per plant tower pot corner or start them in Oasis cubes and transplant one cube with four to six seedlings after 3–4 weeks. It can be harvested for the foliage or the purple flowers for fragrance. It takes 100–110 days from seed to flower. If harvested for foliage begin once the plants reach 8″ in height and continue harvesting as it re-grows to about 10–12″. As the plants age allow longer growth between the crown and cutting height during harvesting to avoid dieback that will eventually occur within the lower part of the plants.

"Sweet marjoram" has an aroma similar to oregano, but sweeter and more balsam-like. It will grow for a year between crop changes providing it is cut back on a frequent basis of every 3 weeks or so. It germinates within 7–14 days and matures within 80–95 days to a height of 8–24″. In the plant towers, directly sow about 10–12 seeds per pot corner. As the plants grow they will hang down as if in hanging baskets. This is similar for most of the herbs grown in plant towers. Begin initial harvesting as the plants reach about 8″ tall. Cut them back lightly to within 3–4″ from their base. Do not allow them to flower in order to keep them succulent. When harvesting any of these herbs that hang down, carefully pull them apart into a bunch with your hand and make one clean cut straight across the top of the bunch keeping at least 4″ of plant remaining. As the plants progressively re-grow allow more of the base to remain to avoid cutting into old material that will eventually suffer some die back.

Spearmint is the most common mint grown for culinary use. Mint seeds are very small so sow, about 15–20 seeds per pot corner. Germination is within 10–14 days. The plants will reach 18–36″ as they mature and hang down in the pots. They mature within 60 days for the first harvest when they reach about 10″ long. As with other herbs in the plant towers, they will hang down and need to be separated by hand during harvesting. These plants will easily re-grow for up to a year providing they are correctly cut back frequently, within 3-week intervals.

"Greek oregano" is used in Italian and Greek cooking. It germinates within 7–14 days and will grow to 8–24″ over 80–90 days. Begin harvesting when it reaches 8″ in length. With time it hangs down and like many of these herbs will completely cover the entire pots of the plant tower (Figure 22.6). Sow about 8–10 seeds per pot corner. Begin harvesting on a regular 3-week period once the plants reach 8″ in length. These plants will continue growing for a year between crop changes.

There are two types of parsleys: curled parsley ("Forest Green") and Italian or flat-leaf parsley ("Giant of Italy"). "Forest Green" is a standard variety of curled or moss parsley. Parsley is slow germinating at 14–30 days. Directly sow 10–12 seeds per pot corner. Parsleys will grow between 16″ and 18″ in height within 75 days. As with other herbs begin harvesting as the plants reach 6″ tall. As they continue maturing, they will hang down in the pots to form a mass of plants. Harvest them regularly every few weeks to keep them succulent. Do not cut back the plants to less than 4–5″ during cropping. They will easily last a year between crop changes. Periodically as dieback occurs in the plant bunches remove the dead plant material to prevent fungal infection. This is also very important with chives. Italian or flat-leaf parsleys have large dark green leaves and strong upright stems initially until they begin to hang down in the plant towers as they mature. If you grow these herbs in pots or beds, they will not lodge as they do in the plant towers.

Rosemary germination is slow and irregular at low rates. To overcome this use "primed" seeds that give faster and more uniform germination than raw seeds. Use primed seeds within 6 months of purchase as viability falls with time. Germination occurs within 2–3 weeks and plants mature between 120 and 180 days. You may start cutting the tips after 80–100 days. It is best to sow (about four seeds) in Oasis cubes and transplant to the plant towers after 4–5 weeks. Starting the seedlings in

FIGURE 22.6   Oregano in plant tower.

Oasis cubes has the advantages of regulating watering carefully and at the time of transplanting allowing the placement of several cubes in a pot corner when only a few seeds germinate in some of the cubes. Rosemary will grow for a year in the plant towers between crop changes. By that time, a lot of woody stems with the dieback of leaves occur in the center of the plants reducing the harvestable succulent shoots. With continued growth, the plants hang down and will have to be cut further out from their base by trimming off the outer 4–5″ of shoot growth.

"Common Sage" has a wide variety of culinary uses. Its dusty green leaves are used in dressings, sauces, and teas and is a great source for dried floral wreaths. Sow six to eight seeds directly in the pot corners. Germination is within 1–3 weeks and maturity at 80–90 days. Start harvesting the shoots when the plants reach 6″ in length. Sage will hang down in the plant towers, so frequent cutting every few weeks is important to keep the plants succulent. Sage can easily last up to a year between crop changes. It will grow from 16″ to 30″ long as it cascades down covering the entire plant tower (if the tower is only sage).

Summer or "Common Savory" with its peppery flavor adds spice to dishes. Sow six to eight seeds directly into the pot corners with germination taking 7–14 days. Plants will grow between 10″ and 18″ within 60–70 days. Start harvesting the shoots when the plants reach 8″ in height. Savory is similar to chervil and will need crop changes every 3 months.

"Summer Thyme" is used in many culinary dishes. It germinates in 3–4 weeks and is ready for first harvest after 3 months. Since it is slow growing, it is best to sow 15–20 seeds in Oasis or rockwool cubes. Transplant to the towers after 5–6 weeks. Sowing in cubes enables the selection and combination of cubes in the plant corners to get adequate plant numbers per pot corner (usually at least 8–10 plants per corner). The plant height is 6–12″ as it hangs down in the towers. Thyme will be productive for about 6–8 months before changing the crop. Start harvesting the shoots when they reach at least 6–8″ in length. Grasp a handful as a bunch and cut straight across with a scissors leaving at least 4″ of growth remaining (Figure 22.7). As the plants mature, they will start to dieback at the center so the shoot tips must be harvested

FIGURE 22.7 Cutting of thyme in "bunch" with scissors.

repeatedly further from the plant base leaving from 6″ to 8″ for re-growth. This will keep them more succulent by not cutting them back hard. If they are cut back hard the plants may die.

When planting herbs in plant towers, it is important as to the position in the tower where specific herbs should be located. Locate upright herbs such as basils and chives in the top pots and the others that hang down in the middle and lower pots. Try to keep the ones that cascade most near the bottom of the plant tower so that they do not grow over the others below. A typical arrangement of a plant tower for growing many of these herbs may be starting from the top pot down: chives, basil (it is preferable to grow basil in NFT to avoid the lodging), parsley, chervil, cilantro, dill, fennel, thyme, lavender, sage, rosemary, marjoram, and mint. Since the number of pots per plant tower should not exceed 10, make up at least two plant towers or combine several herbs in the same pot of the tower.

In summary, when growing herbs in plant towers the plants will hang down as if they were in a hanging basket. This form of growth will cause competition for light for the adjacent pot immediately below. To assist in reducing this shading keep plants pruned from an early stage as soon as they reach 6–8″ in length. Frequent cutting back of the outer shoots will keep the plant succulent and reduce dieback in the center of the plant. When harvesting hanging herbs (as most are) take a handful and bunch them together as you cut the shoots back using a scissors. Be careful to always, especially during the first harvest, allow at least 4″ of growth between the cut and the crown (base) of the plant. Cutting back too hard will retard re-growth and even possibly kill the plants. As the plants mature, continue to cut further away from the crown to avoid cutting into the woody older growth. Do this by cutting a maximum of 4–5″ of the shoot growth at any given time. Do not allow the plants to flower unless their use is for floral arrangements and not culinary additives.

Sow lavender, rosemary, and thyme in Oasis or rockwool cubes and transplant to the plant towers after about 5–6 weeks when the seedlings reach about 3″ tall. Sow the others directly into the substrate of the plant tower pot corners. After sowing, either add about ¼″ layer of medium on top of the seeds or mix the top of the medium so that the seeds go below the surface. It is important to pre-soak the substrate with raw water prior to sowing the seeds and again after sowing. Thereafter, for the first 3–4 days, manually water the pots with raw water using a watering can. Check to be sure that all drip lines are working and that the medium remains moist between irrigation cycles.

Some herbs, such as chives, lavender, marjoram, mint, oregano, parsleys, rose-mary, and sage, will continue growing well for a year provided they are frequently cut back as soon as the shoot tips grow 4–5″ from the previous harvest. Replace thyme after 6 months, and basils, chervil, cilantro, dill, fennel, and savory after 3–4 months. Whenever the plants become dry in the interior of the bunch getting woody, not rapidly forming new succulent growth, they should be changed. Be care-ful with the drip lines: If the stakes become plugged with plant growth then the solution will be reduced causing wilting and dieback of the plants. Similarly, if too much solution is applied and it runs out of the pots moistening the plants, dieback will also occur (Figure 22.8). This dead plant material has to be removed to prevent disease infection. Periodically, prune out dead plant material, especially for chives

FIGURE 22.8   Dieback in sweet marjoram due to excess watering.

and parsley to allow rejuvenation of the remaining shoots from the crown and to reduce the risk of diseases (Figure 22.9).

Several photos are presented to demonstrate the type of growth expected with herbs in plant towers (Figures 22.10 and 22.11).

VINE CROPS

Vine crops include eggplants, European cucumbers, peppers, and tomatoes that are all trained vertically by strings and support wires above. While many staking (indeterminate) varieties may be selected for growing indoors or in greenhouses, it is best to stay with the special greenhouse varieties as they have proven to perform best under controlled environments. Nonetheless, if there are certain heirlooms that you wish to culture for their flavorful fruit, regardless whether they are staking or

FIGURE 22.9   Remove dieback of chives due to aging.

bush (determinate) plants they will grow indoors with hydroponic culture. The main reason for using staking varieties is to better utilize the vertical space in the growing area.

EGGPLANTS

Greenhouse eggplants are becoming very popular in the market due to their smaller size and firm fruit with few seeds and fleshy interior. There are now special greenhouse staking varieties available. High-yielding purple varieties include "Taurus," "Berinda," and "Agora." "Taurus" and "Berinda" are DeRuiter seeds and "Agora" from Rogers NK. "Tango" is a white eggplant with a cylindrical fruit 7″ long by 2″ in diameter. This is also a DeRuiter seed. There are many bush forms of different colors and shapes. These may be grown hydroponically, but they will not yield as heavily

FIGURE 22.10 Location of drip lines in plant towers with thyme. (Courtesy of CuisinArt Golf Resort and Spa, Anguilla.)

as the staking ones. Several I have grown in plant towers are "Hansel" (purple) and "Fairy Tale" (purple and white stripes). These are compact bush types about 2 ft tall. They mature in 65 days compared to the greenhouse staking ones that take about 11 weeks.

EUROPEAN CUCUMBERS

These are the "Long English" seedless (parthenocarpic), burpless types that are shrink wrapped in supermarkets to prevent moisture loss due to their thin skin. There are many varieties; the choice depends largely on your environmental conditions such as light intensity, temperature, and relative humidity. Most are now all female (gynoecious) plants. Varieties I have found to be best for humid conditions that are resistant

FIGURE 22.11   Plant towers (left to right): mint, sage, sweet marjoram, thyme, and chives. (Courtesy of CuisinArt Golf Resort and Spa, Anguilla.)

to Powdery Mildew fungus include "Dominica," "Logica," "Marillo," "Camaro," "Fidelio," and "Flamingo." Other popular cultivars include "Bologna," "Corona," "Discover," "Jessica," "LeReine," "Optima," "Pandorex," "Sandra," "Santo," and "Toska 70." Try a number of different varieties under your specific conditions to determine those most resistant to diseases and most productive. These cucumbers take about 10–12 weeks from sowing to first harvest. In some areas two crops per year are feasible; however, from my experience in many locations, I prefer to change crops every 3–4 months to obtain high yields.

BEIT ALPHA (BA), (PERSIAN PICKLES), (JAPANESE) CUCUMBERS

These are similar to the European cucumbers in that they produce seedless fruits with thin skins. A number of varieties in my experience that are highly productive and tolerant to powdery mildew are "Jawell," "Katrina," "Nimmer," "Sarig," "Suzan," and "Manar." "Manar" is a DeRuiter variety that is most resistant to powdery mildew. Both European and BA cucumber seeds germinate within a few days of sowing. Harvest fruits when the fruit is between 1.25″ and 1.75″ in diameter and from 5″ to 6″ long. The first harvest is usually a week or so earlier than the European cucumbers. Grow at least four crops per year as they cannot be trained the way European cucumber plants are when they reach the support wire. Allow them to grow over the support wire and hang down without cutting the main stem.

PEPPERS

All peppers will grow well hydroponically regardless of whether they are greenhouse or field varieties. However, once again when growing indoors or in a greenhouse use staking varieties in order to use the growing area most efficiently. If you have some favorite varieties that you wish to grow, certainly use them, even if not greenhouse cultivars. With bush varieties, the cropping period will be shorter than the staking greenhouse varieties. Most garden varieties take from 55 to 75 days from sowing to first harvest, whereas the greenhouse ones take about twice that time, normally close to 4 months from sowing. The bush varieties will produce up to several months, so with these cultivars, three crops per year may be rotated, whereas the greenhouse plants yield for 7 months or longer, and so are a single crop per year.

The most popular greenhouse peppers are the sweet bell peppers. Now a few hot peppers and mini bell peppers are entering the market. Bell cultivars are yellow, orange, and red in color. There are many varieties available from various seed houses. Some varieties that I have found very productive include "Bachata," "Cigales," "Lesley," "Samantha," and "Striker." These are yellow fruits, but unfortunately "Samantha" and "Lesley" are no longer available. The seed companies keep on changing varieties, introducing new ones every year, so you must try new cultivars as they are developed. Orange varieties include "Arancia," "Magno," "Orange Glory," "Paramo," and "Sympathy." Red cultivars include "Fantasy," "Jumilla," and "Zamboni." Most of these varieties produce blocky fruits weighing from 200 to 240 g (7–8.5 oz).

Some hot varieties are "Fireflame" and "Habanero Red." "Habanero Red" is a bush (determinate) variety that is relatively hot having a scale of 400,000 Scoville units. The fruit size of "Habanero Red" is 2″ wide by ¾″ long, whereas "Fireflame" is an indeterminate Cayenne type with fruit length up to 6″.

Several mini, small bell, sweet peppers are "Tinkerbell Red" and "Tinkerbell Yellow." These mature in 60 days and produce fruits 1.5″ × 1.5″ (3.7 cm × 3.7 cm).

TOMATOES

Tomatoes fall into a number of types such as beefsteak, tomato-on-vine (TOV), cherry, grape, plum, cocktail, roma, and heirloom. Any of these can be grown hydroponically indoors or in greenhouses, so the choice depends upon personal likes for the specific fruit. All of the varieties exemplified are indeterminate (staking).

Beefsteak tomatoes are the more traditional types of large fruits weighing from 7 to 9 oz (200–250 g). Common greenhouse varieties are "Beverly," "Blitz," "Caiman," "Caramba," "Caruso," "Dombito," "Geronimo," "Match," "Matrix," "Quest," "Rapsodie," "Style," and "Trust." "Blitz" and "Match" are no longer available.

A cluster or TOV tomatoes now make up about 70% of the fresh market. The fruit weight of TOV varieties is between 3 and 5 oz (90–150 g). These are the tomatoes packaged as trusses of fruit, but they can be harvested individually if grown for personal use. Popular red varieties include "Ambiance," "Brillant," "Clarance," "Clermon," "Endeavour," "Grandela," "Success," "Tradiro," and "Tricia." "Lacarno" is yellow and "Orangaro" ("DRK 920") is orange in color.

TABLE 22.1

Recommended Vegetable Varieties for Hydroponic Culture

Vegetable	Type	Varieties
Arugula	Roquette	*Astro
	Sylvetta	Wild Rocket
Bok Choy	Dwarf	*Green Fortune, *Red Choi, *Takuchoy, F1 Hybrid Dwarf Bok Choy, Dwarf Bok Choy, Extra Dwarf Bok Choy, Huo Guo Chai
Lettuce	Bibb/Butterhead	Buttercrunch, *Charles, Cortina, *Deci-Minor, Milou, *Ostinata, *Rex, *Salina, *Vegas
	Leafy	Black Seeded Simpson, *Domineer, *Malice, *Frizella, Waldmann's Dark Green, *Red Sails, *New Red Fire
	Lollo Rossa	*Dark Red Lollo Rossa, Locarno, *Revolution, Soltero
	Multi Leafy	Multi Red 1, Multi Red 2, *Multi Red 3, Multi Green 1, *Multi Green 2, Multi Green 3
	Red Oakleaf	*Navarra, *Oscarde, *Red Salad Bowl, *Ferrari, *Aruba
	Green Oakleaf	*Cocarde, *Tango, *Green Salad Bowl
	Romaine/Cos	Green Forest, *Parris Island, *Outredgeous (red), *Rouge D'Hiver (red)
Lettuce mixes		Encore Lettuce Mix, *Allstar Gourmet Lettuce Mix, *Five Star Greenhouse Lettuce Mix, *Wildfire Lettuce Mix
Basil		*Italian Sweet Basil (Genovese), Aroma 2, *Nufar, *Cinnamon, *Sweet Thai, *Lime, *Mrs. Burns' Lemon, *Spicy Bush, *Pistou, *Dark Opal, *Red Rubin, *Purple Ruffles
Herbs	Chervil	*Vertissimo
	Chives	*Fine Leaf, *Chinese Leeks/Garlic Chives
	Cilantro	*Calypso, *Santo
	Dill	*Bouquet, *Fernleaf, Verling
	Fennel	Bronze, Bronze and Green
	Lavender	*Ellagance Purple
	Marjoram	*Sweet Marjoram
	Spearmint	*Common Mint
	Oregano	*Greek Oregano
	Parsley	*Forest Green (moss or curled), *Giant of Italy (flat or Italian)
	Rosemary	*Primed Rosemary
	Sage	*Common Sage
	Savory	*Common Savory
	Thyme	*Summer Thyme
Vine Crops:		
Eggplants	Greenhouse	*Agora (purple), *Berinda (purple), *Taurus (purple), *Tango (white)
	Garden-Compact	*Hansel (purple), *Fairy Tale (purple and white stripes)
Cucumbers	European	Bologna, Corona, *Camaro, *Discover, *Dominica, Fidelio, Flamingo, Jessica, LeReine, *Logica, *Marillo, Optima, Pandorex, *Sandra, Santo, *Toska 70
	Beit Alpha	Jawell, Katrina, *Nimmer, *Sarig, *Suzan, *Manar

(Continued)

TABLE 22.1 (*Continued*)
Recommended Vegetable Varieties for Hydroponic Culture

Vegetable	Type	Varieties
Peppers	Sweet Bell	*Bachata (yellow), *Cigales (yellow), *Lesley (yellow), *Samantha (yellow), Striker (yellow), *Arancia (orange), *Mango (orange), Orange Glory, *Paramo (orange), *Sympathy (orange), *Fantasy (red), Jumilla (red), Zamboni (red)
	Hot	*Fireflame (cayenne), Habanero Red
	Mini Bell	Tinkerbell Red, Tinkerbell Yellow
Tomatoes	Beefsteak	*Beverly, *Blitz, *Caiman, Caramba, *Caruso, *Dombito, *Geronimo, *Match, Matrix, *Quest, *Rapsodie, *Style, *Trust
	Cluster (TOV)	*Ambiance, *Brillant, *Clarance, *Clermon, *Endeavour, Grandela, *Success, *Tradiro, *Tricia, *Lacarno (yellow), *Orangaro (orange)
	Cherry	*Conchita, *Favorita, *Juanita, *Goldita (yellow), *Zebrino (green stripes)
	Grape/Plum	*Dasher, *Flavorino, *Picolino, *Goldino (yellow), Orangino (orange)
	Cocktail	*Red Delight
	Roma	*Granadero, *Naram, *Savantas
	Heirloom	*Brandywine (red), Striped Green Zebra (green stripes), Yellow Pear (small pear-shaped fruit)

*Varieties that I have successfully grown hydroponically.

Cherry tomatoes have a small fruit size weighing less than 0.5 to 0.9 oz (15–25 g). They also are marketed attached to their flower clusters; however, for backyard growing select the ripest individual fruit during harvesting to get the best flavor. Some popular varieties include "Conchita" (red), "Favorita" (red), "Juanita" (red), "Goldita" (yellow), and "Zebrino" (green stripes).

Grape/plum varieties of similar fruit size to cherry tomatoes are "Dasher," "Flavorino," and "Picolino," which are red; "Goldino," which is yellow; and "Orangino," which is orange.

"Red Delight," a cocktail variety, has a fruit size from 1 to 2.6 oz (30–75 g). This is similar to the Roma varieties that include "Granadero," "Naram," and "Savantas," all of which are red and weigh between 3.5 and 5.3 oz (100–150 g).

Heirloom tomatoes that are indeterminate and therefore suited for greenhouse or indoor hydroponic culture include "Brandywine" (red), "Striped Green Zebra" (green stripes), and "Yellow Pear" (small pear-shaped fruit of ¾–1 oz or 20–30 g).

Most of these tomatoes take 70–80 days from sowing to first harvest. They will also grow for the entire year as a single crop, but if they lose vigor and yields decline they can be replaced once to obtain two crops per year.

Table 22.1 summarizes varieties suitable for indoor hydroponic culture. Of course, as mentioned earlier, you may grow most varieties of vegetable crops hydroponically.

23 Seeding, Transplanting of Vegetable Crops

Like most things in life, if you want to be successful you must start out laying the basis for success at an early stage. Growing plants is similar. You must start with the healthiest, most vigorous seedlings to get healthy, productive plants as they mature. There is a phrase in agriculture that states: "Start with a good seedling to achieve a healthy plant; start with a poor, weak seedling to end in a significantly less productive plant." Optimum environmental conditions will greatly influence the health and productivity of the crop. Important factors that influence seed germination include temperature, water, and oxygen, all of which are important to initiate rapid growth and development of the embryo in the seed to begin a healthy life of the seedling plant. Most seed catalogs give you information on the optimum temperature ranges for seed germination. They often provide preferences of the soil conditions and light for optimum plant growth. Follow those tips closely. Soil conditions generally refer to drainage, which directly influences oxygen levels for seed germination and subsequent growth. The type of substrate plays an important role for the best germination of seeds as it must have adequate moisture retention, but at the same time provide good aeration as oxygen is needed in the germination process. With hydroponic substrates, these oxygen levels will be provided by all substrates, however, for higher moisture retention you can use a peatlite or coco coir medium, whereas more "well drained" substrates refer to vermiculite, coarse sand, and perlite.

Always observe the percentage germination given on seed packages to determine how many extra seeds to sow to get the correct final number of plants. For example, if the germination test indicates 85%, multiply the inverse of that number times the final number of plants (e.g., 30 plants) required (100/85 × 30 plants = 35 plants). It is also a good practice to sow at least 10% more seeds than you need for transplanting as that allows you to select the best seedlings for transplanting and growing on. Using this example then, (35 + 10% × 35 = 39) sow 39 seeds. The number of seeds to sow per growing cube or tray cell varies for different crops, especially herbs that are sown in clusters or bunches. Some seeds, such as lettuce, are available in both raw form and in pelleted form. In some cases, seeds may be purchased as clusters or as primed seed to facilitate sowing and improve germination. In the presentation on individual crops that follows, a guide is given for the number of seeds to sow per cube.

ARUGULA

Arugula seed is fairly small with a seed diameter of about 1 mm (about 1/32″). This small seed is difficult to accurately sow a specific number of seeds. Some seed companies make seed clusters of about 12 seeds stuck together with a water soluble

FIGURE 23.1   Arugula seedlings at 9 days from sowing in rockwool cubes.

adhesive. Usually about 10–12 seeds per cube are the correct density. Alternatively, the seedlings can be thinned upon germination by cutting the extra ones off near their base with a scissors. Be careful not to damage the ones to be kept when thinning. Sow the seeds in Oasis or rockwool cubes (Figure 23.1) or in a peatlite mix in 72-celled compact trays. In pots or beds of peatlite substrate sow the seeds directly into the containers. If you wish to transplant from cubes to nutrient film technique (NFT) or raft culture do so 14–18 days after sowing the seeds.

Optimum temperature range for germination of arugula is 65–75°F (18–24°C) and pH between 6.0 and 6.8. Most vegetable crops like a pH from 6.2 to 6.5. Germination takes from 5 to 7 days. Use raw water for the first 7 days until they reach about an inch high, then use a half-strength nutrient solution until transplanting. They like full light to partial shade. It takes about 3–4 weeks to first harvest and can be harvested two more times thereafter at 3–4-week intervals. As the plants mature they become bitterer in flavor so it is best to replace them after three cuttings.

BOK CHOY

Sow bok choy in rockwool or Oasis cubes and transplant them to NFT, raft culture, or plant towers at 21–25 days. Sow one seed per cube or if more, thin them to one plant at 12 days of age using a scissors when they are about ¾″ high. Geminate the seed with raw water for the first 8 days, then a half-strength nutrient solution until transplanting. Optimum germination temperatures are between 75°F and 83°F (24–28°C). The round, black, bok choy seed is slightly greater than ¹⁄₁₆″ in diameter and therefore is easier to handle than that of arugula.

LETTUCE

Lettuce, as I am sure you are aware, is a cool-season crop, so lower temperatures are more favorable for it. Temperatures from 60–65°F (16–18°C) are best for germination and growth. The lettuce seed can become dormant at high temperatures, especially above 68°F (20°C). Using pelleted lettuce seed expands the temperature range for germination by overcoming dormancy induced by high temperatures. Keep pelleted and raw lettuce seed cool and dry in a refrigerator and use it within a year, especially pelleted seeds. For indoor and greenhouse growing use heat-tolerant varieties that are resistant to bolting (going to seed) and tipburn (browning of outer edges of leaves). Varieties having these qualities are listed earlier in Chapter 22.

Start lettuce in Oasis or rockwool cubes. I prefer rockwool cubes as they do not break apart as easily as Oasis during transplanting, especially in raft culture where it is crucial for the seedling cube to remain intact as the cubes are squeezed a little to fit into the planting holes of the boards. We are fitting square cubes into round holes of the boards. The principle of square pegs in round holes applies here. With some NFT channels, the holes are square and the cubes fit with little effort. The seedlings should be transplanted at 18–21 days of age to their production boards or channels.

Here is a procedure to increase the production efficiency of your growing system by the use of two transplant stages. With a raft culture system use a 2 ft × 4 ft board containing 72 seedlings (six rows of 12 plants) and transplant the first time at 10–12 days as the seedlings reach their third true-leaf stage. Depending upon how many plants you are growing, you can reduce the size of this first-stage transplant to 2 ft × 2 ft with 36 plants. A rule-of-thumb is that the nursery area should be about 20% of the overall system. The seedlings are transplanted to their second-stage (final production) boards at their fourth true-leaf stage (about 18–21 days from sowing) or 8–10 days from the previous transplanting. At this stage there are 18 plants (three rows of six plants) per 2 ft × 4 ft final production board. The lettuce will be ready to harvest within another 24–30 days making the total cycle about 45 days from sowing.

With NFT the procedure is similar. Plants are seeded in Oasis or rockwool cubes (stage 1) and are transplanted for the first time in special nursery NFT channels that have plant holes at 2″ within the channel and channels are separated at 4″ center to center. A 12-ft channel fits 72 transplants. Again, proportion the nursery trays at 20% of the overall production capacity. Start the seedlings in 1″ rockwool cubes (200 cubes/pad) or Oasis 1 "thin cut" Horticubes (276 cubes/pad). Grow these seedlings in the trays on the propagation bench for about 18 days before transplanting to the nursery channel(s) (stage 2). Keep them in the nursery channel for about 10–12 days (28–30 days from sowing) before transplanting them to the finishing channels (stage 3). Within 20–25 days harvest the mature lettuce from the growing (finishing) channels giving you 14 crops per year. In effect, the lettuce only occupies the production area for 20–25 days; that is efficiency in the use of your growing space! Of course, the final production period depends upon the amount of light and day length. This will vary somewhat for a backyard greenhouse due to the season. Using supplementary lights to extend the day length to 14 hours should greatly assist in achieving shorter production periods.

LETTUCE SALAD MIXES

Germinate the seeds as for lettuce between 60°F and 65°F (15.5–18°C). These salad mixes are directly seeded into beds of a peatlite or coco coir mix as outlined in Chapter 22. Harvest within 3 weeks cutting the top shoots with a scissors or electric knife. Allow 3 weeks or less between harvests for re-growth of the leaves. This harvesting can be repeated several times for a total of three harvests between crop changes.

BASIL

Basils germinate very quickly, usually within 4–5 days at temperatures between 75°F and 80°F (24–27°C). Use any well-drained substrate such as a peatlite or coco coir mix if growing them on in peatlite beds or containers. They also germinate and grow well in rockwool and Oasis cubes if you wish to transplant to an NFT or raft culture system. Once the first true leaves appear start using a half-strength nutrient solution to feed them. Transplant seedlings to the final production area after 18–21 days, once they are 2½–3″ tall with at least three sets of true leaves (Figure 23.2). The first

FIGURE 23.2   Sweet basil transplanted at 18–21 days with three sets of true leaves.

harvest should take place approximately 7–10 days after transplanting to force the plant to form multiple lateral shoots as was explained in Chapter 22. I have found that sowing 3–4 seeds in a plug or cube to transplant within 3 weeks gives a good cluster of basil plants. Space them about 6″ × 6″ in the production area. Spacing them similar to the lettuce will permit their developing rapidly into a bushy form of multiple stems. Do not allow them to flower unless you specifically want the flowers as this maturing of the plants causes them to become woody. Regardless, rotate the crop every 3 months when making multiple harvests. Otherwise, for a single harvest, as is usually the case for NFT or raft culture production, remove the plants when they reach 6–8″ in height.

HERBS

Details as to the number of seeds to sow, germination time, and days to maturity of herbs are presented in Chapter 22 when evaluating the best varieties for indoor culture. Best germination temperatures for most herbs lie from 65°F to 75°F (18–24°C). In my experience with growing them in tropical climates, they will tolerate fairly high temperatures, up to 85–90°F (29.5–32°C) under greenhouse conditions. Do not hesitate to grow most herbs under fairly wide temperature ranges, they will survive. Again, I highly recommend planting herbs in plant towers with a peatlite, coco, or perlite mixture as production is greatly increased through training these low-profile plants in the limited space of your home or backyard greenhouse.

With slower to germinate herbs such as thyme and rosemary, start them in rockwool or Oasis cubes and transplant thyme after 6–7 weeks and rosemary at 8–9 weeks. Figure 23.3 shows thyme in Oasis cubes at 30 days that is almost ready to transplant to plant towers.

EGGPLANTS

Eggplants, tomatoes, peppers, and cucumbers are vine crops so seeding and transplanting are similar, with some specific differences. Seed eggplants in 1½″ rockwool cubes. Place one seed per cube. Be sure to thoroughly flush the rockwool cubes with raw water of pH from 6.0 to 6.5 prior to sowing the seeds. This will adjust the pH of the cubes and completely wet the cubes throughout. You do not need to cover the seed holes of the cubes as long as the RH is maintained at least at 75%. Water the cubes several times a day with raw water for the first 8–10 days as the seeds germinate within 6–8 days (compared to tomatoes that take 3–5 days to germinate). Germination temperature is similar to tomatoes at 75–78°F (24–25.5°C). After 10–12 days, replace raw water with a half-strength nutrient solution. Separate the cubes and double the spacing after 10 days laying them on their sides. This side position adds to rooting when transplanting to the rockwool blocks at 14–21 days for tomatoes and eggplants (Figure 23.4).

Eggplants initially are slow in growing, but once transplanted to the rockwool blocks, they begin to grow much faster. Space six blocks per mesh tray flat in a checker-board configuration to maximize distance among the plants (Figure 23.5).

FIGURE 23.3   Thyme seedlings 30 days old in Oasis cubes ready to transplant.

Eggplants are held in the rockwool blocks in trays for a period of 4–5 additional weeks making them 6–7 weeks from seeding before transplanting to the final production system (Figures 23.6 and 23.7). This propagation period is several weeks longer than for tomatoes. Of course, the exact time is dependent upon temperature and light conditions. Be sure to keep the temperature of the medium and nutrient solution at 75°F (24°C) as cool temperatures will slow root growth and overall development resulting in a weak plant. Refer to Figures 23.4 through 23.7 to see the transplanting and subsequent growth of seedlings ready for final transplanting to the growing system.

Various seeding cubes, trays, and transplanting blocks are shown in Figure 23.8.

CUCUMBERS (EUROPEAN AND BEIT ALPHA-BA)

Initial sowing of seeds and propagation is similar for both European and Beit Alpha (BA) cucumbers. Sow one seed per 1½″ rockwool cube after thoroughly moistening the cubes as described for the eggplants. Beit Alpha cucumbers are started and produced under the same conditions as for the European (Dutch) types. Optimum germination temperature is 75°F (24°C). They will germinate within 2 days. Keep

FIGURE 23.4 Laying eggplant seedling on side in cube at 20 days as transplant to rockwool block.

the temperature of the cubes at 73°F (23°C), but do not worry if this is not exact, just maintain temperature within a few degrees of this to get vigorous root growth. If using artificial lighting (indoors) use a 14–18-hour day regulated by a time-clock.

Do not lay the seedlings on their sides as is done with eggplants, tomatoes, and peppers. Just space them to double once the cotyledons (initial seed leaves) are fully expanded and you can see the first set of true leaves forming in the center. Within 7 days, as the first set of true leaves expand to a little longer than the cotyledons, transplant them to the rockwool blocks. Use a dilute half-strength nutrient solution to water them after the first week. Thoroughly flush the rockwool blocks with a half-strength nutrient solution to adjust the pH between 5.8 and 6.2 before transplanting the seedlings to the blocks. Space seedlings at six blocks per mesh tray as for the eggplants. Transplant the seedlings to the final growing area when they are 18–28 days old and have 3–4 true leaves as shown in Figure 23.9. This age depends

FIGURE 23.5 Eggplant seedlings 34 days from sowing and 14 days after transplanting to rockwool blocks spaced six blocks per mesh tray.

FIGURE 23.6 Eggplant seedlings ready to transplant at 47 days old (27 days after transplanting to blocks).

FIGURE 23.7 Eggplants transplanted at 47 days old to bato bucket perlite system. Note: Clamping of stems and drip lines at base. (Courtesy of CuisinArt Golf Resort and Spa, Anguilla.)

upon the temperatures and seasonal sunlight (in a hobby greenhouse) or supplementary lighting indoors.

Cultural techniques for growing BA versus European cucumbers differ in spacing and training, as discussed in Chapter 24.

PEPPERS

Once again, as with the other vine crops, use the same size of rockwool cubes and prepare them with raw water as outlined earlier. Sow one seed per rockwool cube. Peppers like somewhat higher germination temperatures than eggplants, cucumbers, and tomatoes from 77°F to 79°F (25–26°C). Keep these temperatures day and night.

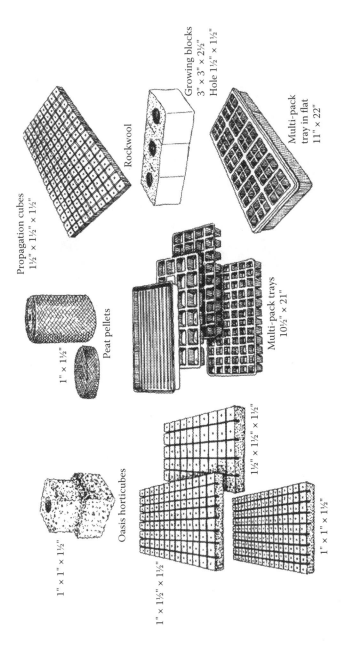

FIGURE 23.8 Various seedling trays, cubes, and blocks. (Drawing courtesy of George Barile, Accurate Art, Inc., Holbrook, New York.)

FIGURE 23.9   European cucumbers ready to transplant to growing system. (Courtesy of CuisinArt Golf Resort and Spa, Anguilla.)

Maintain relative humidity (RH) from 75% to 80%. Once the seeds germinate (about 4–6 days) lower the air temperature to 72–74°F (22–23°C). Provide supplementary lighting of 5500 lux (510 ft-c) with 18-hour day length.

Within 10 days after the first set of true leaves have expanded slightly longer than the cotyledons, lay the peppers on their sides as you double space the cubes in the mesh trays. Peppers grow slower initially than cucumbers and tomatoes. Transplant the seedlings to 3″ rockwool blocks at 3–4 weeks from sowing. Prepare the rockwool blocks with dilute nutrient solution as described earlier for cucumbers and space them six to a mesh tray. When transplanting to the blocks lay the seedlings on their sides in the large round holes of the blocks. By placing your thumb next to the top of the cube while pushing it into the block you can protect the pepper seedling stem from breaking as shown in the photo for the eggplants (Figure 23.4). Air temperatures during this stage should be about 73°F (23°C). The seedlings will be ready to transplant to the final growing area within 3 weeks from putting them in the rockwool blocks (between 6 and 7 weeks from sowing).

TOMATOES

Soak the 1½″ rockwool cubes as with the other vine crops and sow one seed per cube. Use raw water of pH 5.5–6.0 to lower their pH. There is no need to cover the holes of the cubes after sowing. Always adequately water the rockwool cubes and blocks to prevent their drying. Usually several times per day is fine. Regulate temperatures from 77°F to 79°F (25–26°C) day and night during germination. They will germinate within 3 days. Once the cotyledons are fully expanded and the first set of true leaves are 1″ long start using a dilute nutrient solution as was done for the other vine crops. At this stage reduce the temperature regime to 73°F (23°C) during the day and 68°F (20°C) at night.

Lay the tomato seedlings on their sides as they are double spaced in the mesh trays at 10–14 days from sowing. Once the true leaves are fully unfolded (about 3 weeks after sowing) transplant them to the 3″ rockwool blocks being careful to soak the blocks as for the other vine crops and placing the seedlings on their sides into the blocks. Placing them on their sides will promote adventitious roots to form along the buried portion of the stem base as with the peppers and eggplants (Figure 23.4). This increases the number of roots resulting in stronger seedlings. Arrange six blocks per mesh tray in a checker-board pattern as for peppers and egg-plants (Figure 23.5) to give them sufficient space so that their leaves do not overlap as they continue expanding. They will be held in the propagation area for another 3 weeks before transplanting to the production system at an age of 5–6 weeks from sowing of the seeds.

Whenever seedlings are transported to other areas within the propagation location or transferred to new trays, these facilities and growing trays must be sterile. Clean all of these with a 10% bleach solution prior to bringing the seedlings in contact with such items. Likewise, when soaking the rockwool cubes and blocks do it in a clean sink to avoid any contamination. Small precautions like these will go a long way in your successful growing of seedlings.

In soil gardening such cleanliness is not possible; this is a great advantage of hydroponic growing to produce plants free of diseases. It is not difficult and contributes greatly to your success!

24 Training of Vegetable Crops

In Chapter 23, we discussed the details of sowing seeds and the first transplanting stage. For vine crops, it was transplanting the seedlings in their 1½″ rockwool cubes to the 3″ rockwool blocks. The final transplanting to the growing system, at what stage to do this, and subsequent training of the plants through to their maturity are presented here. Procedures vary with specific crops. With all crops at this final transplanting step, you must select the most healthy, vigorous plants first, leaving any smaller, leggy plants last or hopefully to discard them as they should be part of that extra 10% of seedlings that were raised.

With regard to sanitation, there is no point in getting your plants to this final stage of transplanting as clean seedlings and then abandoning those precautions during the final transplantation. That may still introduce diseases. Follow strict sanitation procedures in moving the transplants on carts, in trays, and finally to their placement to the final hydroponic growing system. Keep the seedlings in their original mesh trays until you place them on the slabs or pots ready to be positioned. Do not place seedlings on other nonsterile surfaces before putting them on the slabs or pots. This sounds a little over done, but at any time those seedling roots are exposed to unclean surfaces a disease organism that may be present can enter the roots causing decline in the mature plant as it develops. You have spent a lot of effort in getting the best, most healthy, disease-free seedlings to this final stage so do not jeopardize their continued health through lack of these precautions at the final point.

ARUGULA

There are two ways to cultivate arugula, as mentioned in Chapter 23: either sow directly into beds or pots of peatlite substrate or start them in 1″ rockwool or Oasis cubes and transplant to the hydroponic system at 14–18 days.

In 5-gallon plastic nursery pots sow more or less 30 seeds uniformly on the top. Pre-moisten the substrate prior to seeding and immediately after. With pots a drip irrigation system may be used to water or do so once a day by hand using a watering can.

For beds sow the seeds in rows 6″ apart. Be careful to pre-wet the substrate before sowing the seeds. If you want to automate the irrigation of beds place "ooze" or "soaker" hoses one foot apart along the length of the bed. Soaker hoses are thin-walled drip lines of ½″ diameter that have very small holes every 4″ along the hose. This gives more uniform distribution of the water along the bed length. You can make your own "soaker" hose by using ½″ black poly irrigation tubing and inserting 0.5 gph emitters into it at 4″ centers. Do not place drip lines on the emitters, simply let them drip directly below to the substrate. At 3″ on each side of the irrigation line sow the seeds in line. This can easily be done by using a broom handle or ¾″ diameter pipe to press a groove into the substrate ½″ deep. Sow the seeds into this channel almost touching each other about three to four seeds across. This will give a dense growth of the arugula in the rows and they will spread to the side forming a complete cover of plants in the entire bed. From sowing the seeds to the first cutting, it will be about 4 weeks. When harvesting use a scissors or electric knife to cut sections of the bed at a time. After three harvests, about 3 months, it is best to remove the plants and seed again. If you want continued production, seed half of the bed length every 8 weeks or one-third of it every 5–6 weeks to keep continuous young plants growing that will replace the older ones after their third harvest.

For transplanting to nutrient film technique (NFT) or raft culture systems start the seedlings in rockwool or Oasis cubes. Place 10–12 seeds in the each cube after the cubes have been thoroughly flushed with raw water to adjust their pH to near 6.0. To increase the efficiency of production, as was described in Chapter 23 under "Lettuce," use two transplant stages. Arugula has the same spacing and a similar growth rate to lettuce, so follow the lettuce procedure. The first transplant stage is at 10–12 days when the seedlings have formed three true leaves. With raft culture transplant them to the nursery raft having 4″ × 4″ spacing. In NFT transplant them to the nursery channel having holes at 2″ centers with channels spaced 4″ center-to-center. They are held in this nursery location for 8–10 days before transplanting them to the final finishing or production boards/channels. They will be ready to begin harvesting within 2 weeks.

If you are not growing a lot of arugula, this extra transplanting stage probably does not justify the additional work. Then, simply transplant to the final production boards/channels at 12–18 days from sowing in the cubes. It is more feasible to use the two-transplant procedure if you wish to harvest the arugula one time at its maturity between 3 and 4 weeks and not make multiple harvests.

When the seedlings are transplanted to the production area use a complete nutrient solution, not the diluted half-strength one used on the seedlings when on the propagation bench.

BASIL

The cultivation of basil is very close to that of the arugula described earlier, except that it should be transplanted and not directly seeded. If you wish to transplant to pots or beds of peatlite medium, start the basil seeds in the same peatlite mix in 72-celled compact trays. Sow four to five seeds per cell or cube as outlined in Chapter 22. For NFT or raft culture follow the preceding procedures for arugula. The cropping cycle closely follows that of arugula.

The principal difference between arugula and basil is the harvesting method. Arugula is simply cut back to within 2–3″ above the crown area, whereas basil must be trained through precise cutting stages. The first cut of basil is as it reaches three sets of true leaves at the third- to fourth-node. Cut the tops about ¼″ above the node of the leaves. You may think that this is more complicated than growing in soil, but the fact is the procedure is exactly the same whether growing basil hydroponically or in soil if you wish to maintain a juvenile plant to prevent it from becoming "woody" stemmed. Subsequent harvests take place as the plants produce side shoots with at least three sets of leaves. Repeat this cutting technique with all side shoots as they continue to bifurcate as they are cut back. This produces bushy (multi-stemmed) young growth. However, after about 3 months the plants need to be replaced as they eventually become woody as they mature, especially if you allow them to flower. Seed the new crop about 3 weeks before taking out the older plants so that the down time between crops is reduced to 10–12 days. You may also sequence the cropping so that a portion of the plants are at different ages. In that way, you get continuous production.

Feed the plants with a complete nutrient solution after transplanting to the production system. Optimum temperatures for basil are from 70°F to 80°F (21–27°C).

BOK CHOY

Bok choy is transplanted to plant towers, NFT, or raft culture 21–25 days after sowing in rockwool or Oasis cubes. To transplant to the plant towers place one seedling in each corner of the pots. Use the end of a broom handle to indent a hole in the substrate to a depth of 1½" to facilitate the placement of the seedling with its cube. The crown (area where the plant stem enters the cube) of the plant should be at the surface of the medium. Water the substrate before and immediately after transplanting. You may have to water a few times by hand over the next few days until the roots take hold in the substrate to prevent any possible drying of the seedling cube. After that the drip irrigation system will provide sufficient moisture with a full-strength nutrient solution. Bok choy grows well at temperatures from 70°F to 80°F (21–27°C). The first harvest is 20–25 days after transplanting when a semi-head forms. Of course, you may harvest it at earlier stages, particularly if you want tenderer "baby" bok choy.

If you are growing bok choy in one plant tower, sequence the planting of each pot a few days apart to obtain continued production. For example, if the plant tower has 10 pots, each pot will have four plants, except the top pot that can have eight plants to give a total of 44 plants. If you sow four plants every second or third day you will get plants ready to harvest every day. Unless you are really crazy about bok choy, this one plant tower will be too much for your personal consumption, so why not use half of the plant tower for bok choy and the other half for arugula, basil, or lettuce!

To grow the bok choy in NFT or raft culture, the cropping cycle is the same as for that of the plant towers. You will not have to hand water after transplanting as they receive solution during the continuous irrigation cycles of these hydroponic methods. I have grown it in the A-frame NFT system to increase production (Figure 24.1).

LETTUCE

The normal single transplant procedure at 18–21 days to the production system (Figure 24.2) may be used or if you wish more efficient utilization of space follow the two-transplant method. The more efficient method of several transplantings into

FIGURE 24.1 Bok choy in NFT A-frame. (Courtesy of CuisinArt Golf Resort and Spa, Anguilla.)

FIGURE 24.2 Lettuce seedlings 20 days old in rockwool cubes.

raft culture and NFT were described in detail in Chapter 23. Here is a brief summary of the cropping procedure:

1. Sow in rockwool or Oasis cubes.
2. Grow for 18 days (Stage 1).
3. Transplant to the nursery board/channel at 18 days (Stage 2).
4. Grow for 10–12 days (28–30 days from sowing).
5. Transplant to production (finishing) board/channel (Stage 3).
6. Grow for 20–25 days in the production system.

Total days from sowing to harvest: $18 + 12 + 25 = 55$ days. This gives 14 crops per year. Remember that the cycle length will vary by a number of days depending upon your light source. In a backyard greenhouse, the cycle may be reduced by up to 3–5 days during the long day length of summer and increased by 5 days or more during the short winter days. Supplementary lighting of 14-h day length will assist in reducing this period during the winter months.

Since you are growing this lettuce for your personal use, it can be harvested earlier, especially if you wish a "baby" form that is just over half the normal mature age. Most folks like salads every day, so in that case the cropping cycles are sequenced to meet that objective. For two lettuces per day, it is easiest to sow four to five every second day and then harvest a few heads several days younger than others. This is a little more work in that every few days you must sow and transplant to replace the ones harvested.

Ideal temperatures for lettuce are from 60°F to 75°F (15.5–24°C). Use a full-strength lettuce formulation upon transplanting. You should be able to purchase a leafy salad formulation from a hydroponic outlet. This formulation is good for lettuce, arugula, basil, and bok choy. If such a formulation is not available, you can always use a general vegetable one.

HERBS

Most efficient growing of herbs is in plant towers. They can also be grown in pots or a bed of peatlite medium by direct sowing. Some herbs such as chives, mint, thyme, and watercress do fine in NFT and raft cultures. However, when using these methods sow the seeds in rockwool or Oasis cubes and transplant. The details of growing herbs were presented in Chapter 22, so please refer to that section. Most herbs tolerate wide temperature ranges. Use a leafy vegetable formulation or any other available.

EGGPLANTS

Eggplants started in rockwool cubes and blocks are transplanted to the hydroponic production system at 6–8 weeks of age as shown in Figure 23.6 of the previous chapter. They will have about three pairs of leaves at that stage.

It is important to prepare the final growing system prior to transplanting. This includes securing the support strings, soaking and flushing of the rockwool or coco coir slabs and perlite in pots. Support string should be wound onto "Tomahooks," special hooks containing extra string for the lowering of vine crops (Figure 24.3). These hooks can be purchased from CropKing or other greenhouse suppliers or make up your own with galvanized #9 wire as shown in the diagram. Put about 40 wraps of string on the hook in order to accommodate lowering of the plants. The hooks allow you to move the plants at the top and lower them in a process called "lowering and leaning." As you move the vine crops in one direction to take up the stem length, you lower them by unwinding several rounds of string. Use plastic string as it is strong and remains clean, not accumulating dust or fungal spores. As the seedlings are placed on the slabs or pots fasten a plant clip under a pair of strong leaves (Figure 24.4) to support them by the string. The clips have a hinge at the back that pinches the string to secure it from sliding down.

Rockwool slabs must be thoroughly wetted prior to placing them out. You may do this by irrigating the slabs until they fill up with solution (normal strength at this time) as they have no drainage slits made as yet. Let them soak for about 24 h. Then, assuming you have already made the collection tray for the slabs to sit in and drain, cut 1″ slits on the inner side in three places that are not directly under the plant locations. This will permit drainage and collection of the leachate by the

FIGURE 24.3 Tomahooks attached to overhead support cable. (Courtesy of CuisinArt Golf Resort and Spa, Anguilla.)

FIGURE 24.4 Placement of plant clip on tomato plants during transplanting. (Courtesy of CuisinArt Golf Resort and Spa, Anguilla.)

underneath gutter to direct the solution back to the nutrient tank. At the sites where the plants will be placed cut a cross (X) wide enough to fit the rockwool block as you place it directly onto the rockwool medium. Insert a drip line stake into the corner of the block.

If using coco coir slabs, the procedure is similar in soaking the slabs, but they will already have drainage holes. Flush them to get complete moistening of the medium and allow at least 25% leachate with this initial irrigation. Thereafter, irrigate to 15% leachate. The nutrient solution will adjust the pH and electrical conductivity (EC) of the slabs to 6.0–6.4 and 2.0 mMhos or slightly higher depending upon the nutrient formulation.

Plant spacing with both of these types of slabs is the same by placing one row of slabs end-to-end with five plants in each slab. The slabs are slightly longer than 3 ft (1 m or 39″). Single rows are 6 ft apart. Position the plants on the slabs to get them equally distant between all plants on the slabs. The first plant and last plant in each slab is about 3″ from the ends with the others 8″ apart since the slabs are actually 39″ long. You do not need to be exact. As the plants grow, the support training above will space them equally. The spacing should work out close to 4 sq ft per plant.

Bato buckets with perlite substrate are spaced a little differently as discussed in Chapters 13 and 20. Bato buckets are located on top of a 1½–2″ PVC drain pipe at 16″ centers. The pots are staggered on each side of the drain pipe. The individual plant area approaches 4 sq ft as with slabs. The difference is that each pot contains two plants (eggplants, tomatoes, peppers, Beit Alpha (BA) cucumbers) or one European cucumber (8–9 sq ft/plant).

Pre-soak the perlite with raw water and a complete nutrient solution prior to transplanting. During transplanting bury the seedling blocks about three-quarters of the way into the perlite. With the support strings already in place fasten a plant clamp under a pair of strong leaves to keep the plant upright and prevent it from lodging that may break the plant. Secure one drip stake in each block at the corner, not immediately adjacent to the crown of the seedling.

The foregoing is the transplanting procedure, and now we must look at the training techniques for the continued growth and development of the eggplants. They, like peppers, are trained to two stems per plant. Some growers may keep three stems per plant, especially if the plants are spaced further apart to get more than 4 sq ft per plant, that is, at lower plant density. I have found that two stems with plant area at 4 sq ft works well. The main stems of eggplants are not flexible like those of tomatoes so when they reach the support cable above, remove the growing point.

FIGURE 24.5   Pruning side shoot of eggplant. (Courtesy of CuisinArt Golf Resort and Spa, Anguilla.)

FIGURE 24.6 Clamping eggplant.

Keeping more stems also slows the growth rate of the plants in arriving at the wire. For shorter crops (less than 5 months from seeding), it is better to increase the planting density with less stems to keep vigor.

As the plants reach about 2 months from sowing of the seeds or about 2 ft in height select the two most vigorous stems to continue developing and prune out the others as shown in Figure 24.5. Remove all side shoots up to 3 ft in height and then cut back side shoots to one node. Clamp the main stems using separate support string for each shoot. Attach plant clips under sturdy leaves about every foot along the main shoots (Figure 24.6) or twist the string around the stems for support. Later when the plants are 6 ft tall, allow additional side shoots to form beyond three to four nodes to get extra production and slow down the crop as it approaches the support wire. Sow seeds 6–8 weeks before removing the existing crop so that they will be ready to transplant immediately into the production system when the old crop is terminated. You may harvest all the eggplants, including small "baby" ones, as you take out the plants.

Eggplants do not form large clusters of flowers as do tomatoes. The flowers that form are variable in vigor, so select the large strongest flowers to remain and remove other smaller ones. This will assist in obtaining large, more uniform fruit. You may pollinate the flowers with an electric toothbrush as for tomatoes, but it is not necessary. Commercial growers often use a hormone weekly to create more fruit development, greater fruit weight, and earlier production. You may harvest the fruit when it changes in color from a dark, shiny purple to a little more pale purple at the tip of the fruit and becomes slightly less glossy. Use a pruning shears or scissors to cut the fruit from the plant to get a nice, clean cut that will heal rapidly, thus decreasing any threat of diseases.

Optimum temperatures for eggplants are 70°F–72°F (21–22°C) during the day and 65°F–68°F (18–20°C) at night. A tomato nutrient formulation suits eggplants fairly well, but they are more susceptible to Mg deficiency indicated by yellowing of the older lower leaves. A nutrient solution EC of 2.5 mMho is good with pH 6.0–6.5. Enrichment of the atmosphere with CO_2 (carbon dioxide) at levels from 700 to 1000 ppm will increase production, but do not exceed 1000 ppm.

Eggplants are a great crop to grow. They produce beautiful flowers and deep, dark purple fruits (purple varieties) or snow white fruits with white varieties. If you like to cook with eggplants, growing your own is very rewarding in that you get firm fruits with few seeds in the greenhouse varieties. In my experience, they are one of the easier vine crops to grow and are very impressive in appearance. Try them, you will be happy to have done so!

PEPPERS

Peppers are next as they are trained similar to the eggplants. The cropping cycle of peppers is based on a single crop annually. They take 4 months from seed to first pick; therefore, more than one crop per year is not feasible, unless they decline greatly in productivity. These greenhouse varieties of peppers grow to about 14 ft in height over this yearly cycle. In low hobby greenhouses or in your home, you will have to lower them, similar to tomatoes. Since pepper stems are brittle, you must exercise caution in lowering them slowly and moving them down the support cable as is described later. You can also grow the normal bush type of garden pepper, if you do not want to deal with the extra work of lowering the staking peppers.

Peppers are transplanted to the production system at 6 weeks of age. At this age, they should be about 10″ tall with about four leaves on the main stem. At this stage, some roots will have begun to emerge from the base of the rockwool block.

Once again prepare the slabs or pots of substrate as presented for the eggplants. Wet the slabs or substrate in pots with a nutrient solution for 24 h prior to setting the seedlings onto the medium. Install the tomahooks with string to the overhead support cable. Transplanting is the same as for eggplants and tomatoes. Spacing and plant density is also the same. Peppers, like eggplants, are trained to two main stems. Support the plants with plant clips on the string when transplanting.

Optimum day temperature is 70°F (21°C) and night temperatures of 61°F–63°F (16–17°C). Maintain a feed EC of 2.5–3.5 mMhos with pH 6.0–6.5.

Within 2 weeks of transplanting, the plants will form two to three stems at a fork. Prune out the third stem if developed leaving the two strongest stems. Each stem must be secured with plant clips to the support string. If the slabs or pots are arranged in a single row, use a V-cordon form of training the plants to the overhead wire where every other plant is supported to the opposite overhead wire as shown in the diagram (Figure 24.7). As the plants grow, they develop side shoots at each node (point on the stem where leaves form). These shoots must be pruned back to allow

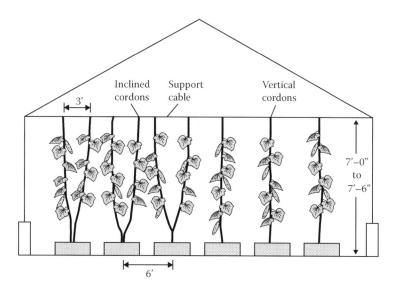

FIGURE 24.7 V-cordon training of vine crops. (Drawing courtesy of George Barile, Accurate Art, Inc., Holbrook, New York.)

one node with a leaf. This extra leaf assists in increasing the overall leaf area of the plant that should help in getting larger fruit. Prune the side shoots by pinching the growing point with your fingers or if they get large use a scissors or pruning shears. Cut them close to the leaf and shoot node.

It is best to remove the first flower that forms at the fork and let the first flower to set fruit at the second node above the fork. This keeps the plant more vigorous initially before bearing fruit. You may also wish to remove the flower at the third node, especially if the plant is showing lack of vigor. After the fourth flower allow all the flowers to set fruit. Unlike tomatoes, pollination of the pepper flowers occurs without assistance from bees or a vibrator such as an electric toothbrush. However, a few cultivars may respond to pollination by enhancing fruit development.

Harvest the fruit with a sharp knife or pruning shears when the fruit is at least 85% colored. Be careful not to cut into other fruits or the stem when cutting the fruit stem (peduncle) of the pepper. The plant must not develop too many fruits at one time as that will reduce the size of the fruit and restrict the growth of the plant. A good balance is to maintain five to six developing fruits per stem. If there are more fruits set, remove the smallest ones among the larger fruits early in their formation.

Color peppers, red, yellow, and orange, are so much more flavorful and nutritious than the green, unripe stage of the fruit. One thing about colored peppers, regardless of whether they are grown in the field or greenhouse, is that they must be 85% vine ripened before harvested as they cannot be picked green and gassed with ethylene to change color as is the case with tomatoes. That is why when purchasing colored peppers in the supermarket they have a superior sweet taste. Growing your own will take that quality one step further by allowing the fruit to ripen 100% in order to gain the best flavor and nutrition. This is a great advantage in growing your own vegetables!

TOMATOES

Tomatoes are the most popular vegetable grown for fresh salads. By growing them yourself, you will benefit from superior flavor and nutrition compared to supermarket products as you can completely vine ripen the fruit. It takes about a week after the fruit begins to change color from green to become fully ripe. Most tomatoes, even greenhouse hydroponic ones, are picked just when they begin to turn color and are therefore about a week away from full ripeness. Of course, they must be harvested at this firm stage to enable shipping them for long distances.

Because of this immature ripeness and gassing the fruit with ethylene upon arrival or in route to the market, the tomatoes have almost no taste compared to fully vine-ripened ones. This excellent flavor and nutrition obtained by fully vine ripening the fruit is your reward for your effort and investment in growing your own tomatoes. And it is fun as a hobby, in addition to the benefits of self-satisfaction of producing vegetables, their very distinct "real" taste and nutrition, not otherwise possible with store-bought produce, will keep you enthusiastic about growing with hydroponics!

Initial training of tomatoes in transplanting at 5–6 weeks from seeding is similar to the eggplants and peppers. Prepare the slabs, bato buckets, pots, and so on as was done with the eggplants and peppers. Have all of the plant support system in place before transplanting. A very healthy tomato transplant is said to be "as wide as tall" at its third to fourth pair of leaves when ready to transplant.

Spacing of tomatoes is the same as for eggplants and peppers at about 4 sq ft of greenhouse floor area per plant. Arrange five plants per slab or two plants per bato bucket as described for eggplants. Clamp the young plants under a strong leaf to support strings with a plant clip to the support string. Training from there differs from the eggplants and peppers in that the tomatoes are trained to one stem only.

Remove all side shoots that form between the main stem and each leaf axil (where the leaf joins the stem). Removal of the "suckers" takes place several times a week in order to pull them off by hand while they are still small, best as they reach 1″ in length (Figure 24.8). Grasp the main stem just below the leaf axil with your thumb and index finger pressing lightly the stem and leaf base so that as you quickly pull the sucker off with the other hand, you will get a clean break as shown in Figure 24.9. A clean scar will heal rapidly without any threat of disease infection. If the suckers get to 2″ or longer, you should use a pruning shears or sharp knife to cut them off close to the base at the leaf axil. Wear vinyl disposable gloves when working on the tomatoes as the acid of the plants causes cracking of your skin and is hard to remove from your hands.

Pollination is very important in achieving high "fruit set," which is the fertilization of the flower and subsequent development of the fruit. Fruit set is successful as you see the small beads of fruits forming (see Figure 24.10) after the blossoms fall. Tomato flowers remain receptive (petals bend back as shown in Figure 24.11) for several days making pollination a daily task. Normally, in commercial greenhouses bumblebees are introduced into the greenhouse to pollinate the flowers. This is not feasible on a small scale, so the best alternative is to use an electric toothbrush. The flowers are termed "perfect" in that they have male (anthers) and female parts (pistil). With the high vibration of an electric toothbrush, as you place the bristle end at

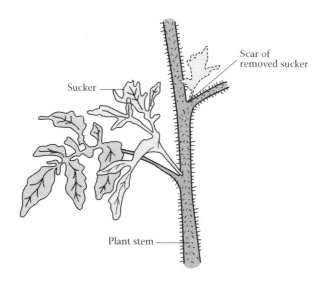

FIGURE 24.8   Tomato sucker (side shoot). (Drawing courtesy of George Barile, Accurate Art, Inc., Holbrook, New York.)

FIGURE 24.9   Breaking sucker off with position of thumb grasping main stem.

FIGURE 24.10 Fruit set of tomato.

FIGURE 24.11 Receptive tomato flowers.

the flower stem (cluster or truss), it will dispel the pollen from the anthers and cause some of it to attach to the female (pistil stigma).

In the case of beefsteak tomatoes, if you wish large fruit, thinning of the tomato fruit on the clusters (trusses) to four to five fruits will produce higher fruit weight. If thinning, do this shortly after fruit set while the fruits are still small (about ¼″ in diameter), so that energy and food is not going to the fruit that you do not want to develop. If your plants are "vegetative," that is, have a lot of large leaves and few fruit trusses, do not take off any fruit from the trusses as keeping all the fruits will help to shift the plant to a more productive state ("generative"). Cherry and "Tomato-on-Vine" varieties do not require truss fruit pruning.

As the tomato vines grow tall and approach the overhead support wire, you will have to "lower and lean" the plants. Do this as you are suckering the plants when they reach the wire above. Do not allow the plants to grow beyond the support wire and fall over as that will cause breaking of the growing point and termination of your crop. Generally, you will have to lower the plants every week about a foot in length at a time to prevent any breakage and lessen stress on the plants. Lower the plants by unwinding several wraps of the string on the "tomahooks" and moving the hook along the support wire in one direction. During this process, the lower portion of the stem may be laid down. Support the lower stem portion by a wire support above the slab or rest it on top of the bato buckets or pots. Stem supports may be made of ¾″ or 1″ diameter PVC pipe with tees or using number nine galvanized wire (Figures 24.12 and 24.13). Alternatively, you can make a PVC support frame above the slabs to hold the lowered plant stems. Then, remove three to four leaves from the base of the lowered main stem to improve air circulation, thus avoiding potential disease infection. Do not remove more than three to four leaves at any time during a week or you may shock the plant by reducing too much leaf area at a time.

You may think this is a lot of work. It is, but recall that in your backyard garden you do not have to follow such a procedure as your tomatoes will get frozen out before they reach beyond 7–8 ft in height. However, even outside with a summer crop, as the tomatoes are harvested the lower leaves will yellow and fall off. To prevent fungal infection, it is best to remove these leaves as you would for indoor hydroponic culture.

Within 4–5 weeks prior to starting a new crop, pinch the growing tip of the plants. This will stop any further vegetative growth and force nutrients to the remaining fruit as they ripen. During the crop change over remove all plants, clean up all plant debris, change the slabs or medium in pots, and sterilize the pots, floor, nutrient tank, drip lines, return channels, tomahooks with string, and so on with a 10% bleach solution. Soak the components in the bleach solution for at least several hours. This sterilization process is the key to preventing any disease or insect carry over to your next crop. You will then be ready for a new start with the next crop. Start your seedlings in another location under supplementary lights 5 weeks prior to removing your crop. In that way, as soon as you have discarded all the old plants, cleaned and sterilized the area and growing components, you can immediately transplant the new crop to the hydroponic system. This clean-up applies to all crops during their change to a new crop.

The normal cropping cycle for tomatoes is 1 per year. If you find your crop slowing greatly in productivity, you can terminate it and use a two-crop system annually.

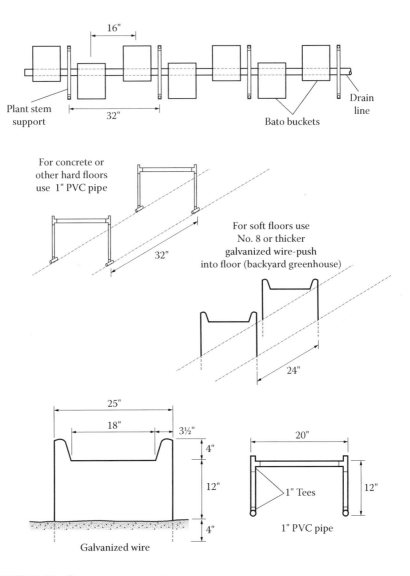

FIGURE 24.12 Tomato stem supports.

This may be necessary due to pest or disease infection or simply restricted space to accommodate the plants as they continually need lowering.

CUCUMBERS (EUROPEAN AND BA)

I want to discuss these last as their training is somewhat different from the other vine crops. European cucumbers are transplanted to the growing system as the second set of true leaves initially form, but are not fully expanded. This is generally from 14 to 21 days from sowing. The exact time is dependent upon light and temperature

FIGURE 24.13 Tomato pipe stem supports. (Courtesy of McWethy Farms, Three Oaks, Michigan.)

conditions. Do not allow them to grow much beyond this stage as they will fall over (lodge) and it will be difficult to transplant them without breaking their stems. Healthy seedlings ready to transplant are shown in Figure 23.9.

European cucumbers are spaced at half the density of tomatoes (8–10 sq ft per plant). Secure the plants while transplanting to the support strings with the plant clips. Place one plant per pot or two plants per slab. Locate one drip line at the corner of the rockwool block for each plant. In pots having one block with its seedling, place one drip stake at the corner of the block and the other in the substrate of the pot. Within 4–5 days as the plants spread their roots into the substrate and become established, they will grow 6–8″ per day. This means you must train them every day. As they grow support them by wrapping the support string one time between each set of leaves (around the internode). The best way to accomplish this without breaking the tender stems is to pull down on the support string allowing some slack and wrap in a clockwise direction so that you remember not to unwind the plants by not knowing which direction you wound the string the previous day. This is a small but important detail to efficiently care for the plants as it is a daily chore, with pleasure of course, as you see them growing so fast!

You must also remove all the tendrils (the stringy appendages they produce seeking support) on a daily basis. The best procedure then for daily training is in the order of removing the tendril(s), wrapping the string, and then removing side shoots up to several nodes below the tip. By pinching the side shoots last, you have the opportunity of keeping the plant growing should you break the main stem (leader) in the process of wrapping the string. If breakage occurs, simply allow the nearest side shoot to grow replacing the original growing point. Finally, remove all small fruit up to six nodes to give the plant a chance to grow large leaves that later will be able to support the developing fruit on the main stem.

When the main stem arrives near the support wire above, do not pinch the last two to three side shoots as they will develop as two laterals, one hanging down on either side of the main stem. Support one side shoot on each side of the main stem to the support wire with a plant clip and allow it to grow down as shown in Figures 24.14 and 24.15. Pinch side shoots from these laterals below the third one so that as the fruit is harvested from the laterals (usually only two to three) the next set of side shoots will replace the first as you cut off the older side shoot once it has finished bearing fruit. If you have lots of light you may try leaving several side shoots to develop above the main lateral as they pause in growth until the fruit is harvested from the main lateral.

Pollination is unnecessary as the flowers are all female. That is evident from the miniature fruit behind each flower. You do not want seeds in the cucumbers, they are seedless, so no pollination! If by chance any male flowers form, take them off to avoid pollination. The male flower has no small fruit behind it as shown in Figure 24.16. ·

Keep the cropping cycle for European cucumber indoors short at 3 months from seeding to termination. At that time, two to three laterals will have produced fruit.

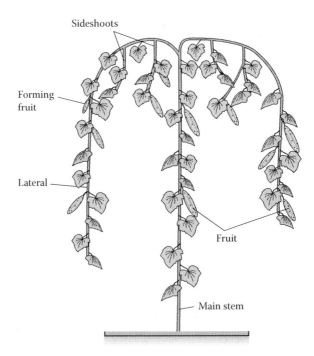

FIGURE 24.14 Training of top of cucumber to allow two laterals over support wire. (Drawing courtesy of George Barile, Accurate Art, Inc., Holbrook, New York.)

FIGURE 24.15 Training of cucumber over support wire. (Courtesy of CuisinArt Golf Resort and Spa, Anguilla.)

FIGURE 24.16 Male cucumber flower.

As the plants get older they become less productive and more prone to pest and disease issues, so to avoid this simply plan on a 3-month crop giving you four crops annually. Remember to start the new plants several weeks before removing the old crop so that as soon as the old crop is taken out you can transplant to the new substrate in pots or slabs. You may attempt to sterilize rockwool or coco coir slabs between crops, but that is a pain as you must remove the plastic wrap and heat them in the oven of your stove to 180–200°F (82–93°C) for at least half an hour. After sterilizing the slabs with heat, wrap them with new polyethylene. In the case of pots, sterilize the pots for at least 1 hour in a 10% bleach solution, let them dry, and then fill them again with fresh medium.

European cucumbers should bear two to three fruits per week per plant, but they stop bearing for about a week as the fruit is all harvested from the main stem and the laterals begin to develop. The laterals do not grow a lot until the main stem fruit is finished.

European cucumbers will grow on a tomato or general vegetable nutrient formulation of EC 2.5–2.8 mMho and pH 5.8–6.3. To avoid the details of making up your own formulation purchase a premixed formulation from a hydroponic outlet. When growing these plants indoors, I recommend that you regulate the environment of temperature and humidity to most closely conform to that of tomatoes as they will likely be your main crop. When growing many crops together, you cannot expect to satisfy all conditions at optimum levels for each one as that is not feasible. Under good conditions for tomatoes, you can successfully grow the other vine crops. Lettuce, however, prefers a lower temperature regime than the vine crops. Keep the lettuce system close to floor level where temperatures are likely to be cooler.

BA cucumbers are spaced and trained differently from European cucumbers. BA cucumbers are spaced similar to tomatoes, about 4–5 sq ft per plant. Transplant four plants per slab or two plants per pot. Transplant these cucumbers at the same two-leaf stage as for the European ones. Secure each plant to a support string using a plant clip and position a drip line at the corner of each rockwool block.

The subsequent training differs in that only the first six to eight side shoots are removed on the main stem and thereafter are pinched at the second node. These cucumbers set multiple fruits at each node. As the plants reach the overhead support cable do not cut the main stem, let it grow over the wire and hang down (Figure 24.17). With a plant clip attach the stem to the overhead support cable. The main stem will continue to grow with its side shoots. Prune all tendrils to prevent their tangling up among other shoots. As the main stem grows down cut the side shoots at three to four nodes. Remove any deformed fruit early to maintain continued production. If the plant sets excessive fruit many will abort. BA cucumbers like the European ones are gynoecious (producing only female flowers). Harvest the fruit at 1.5″ in diameter by 5–7″ in length (depends on the specific variety).

BA cultivars produce 2–3 times as many fruits as European cucumbers, but of course, they weigh a lot less at about 4–6 ounces compared to European types weighing 14–16 ounces. The cropping cycle should be about 3 months to get four crops per year. The EC of the nutrient solution should be maintained between 1.5–2.5 mMhos and pH 5.8–6.5. Use a general hydroponic vegetable formulation available

FIGURE 24.17 Beit Alpha cucumber training main stem over support cable. (Courtesy of CuisinArt Golf Resort and Spa, Anguilla.)

from hydroponic suppliers. Other environmental conditions favorable for European cucumbers are good for BA cucumbers.

With regard to flavor, I find the BA cucumbers have a more distinct "cucumber" flavor than the European ones. One reason they are termed "Persian Pickles" is that the fruit is smaller in diameter and denser giving them a crisp texture. Because of this crispness, they are very sought after by the Middle Eastern ethnic groups as they can be pickled and not just eaten fresh as are the European cultivars (cultivated varieties). Try pickling some and compare them to the traditional garden types. Certainly, they will have softer skins, yet still retain the crispness internally. They are a great treat for salads and pickling!

25 Pest and Disease Control

You may think that growing indoors or in a greenhouse will exclude all pests from getting into your crops. That is not the case; the insects always find their way into any crop location. Even in mid-winter when a lot of the insects are killed by frost, those that found their way into your crop earlier will thrive unless you control them. Environmental conditions within your indoor or greenhouse garden are ideal for pests. Diseases, on the other hand, can be controlled by use of resistant varieties and keeping moisture levels down. In general, a relative humidity of 75% is good for your crops and is not too humid to promote diseases. However, be careful to avoid any leakage from the growing systems that will build up moisture levels and at the same time become a home for algae. As soon as algae grow in wet spots, they attract insects like fungus gnats that will feed also on plant roots, especially of your seedlings. Clean up any leaks and keep the floor dry to prevent these issues.

CLEAN-UP BETWEEN CROP CYCLES

Whenever the crop is terminated and all plants disposed of, the entire growing area must be sanitized to prevent carryover of pests and diseases to the next crop. Prior to pulling the crops spray them with a pesticide to kill most of the flying insects. As you remove the plants place them in large garbage bags to prevent infested plants from spreading pests and diseases. The plant debris can be taken to a garbage land fill or you can bury them in a pit outside covering it after with soil. This is the more difficult way as the pit would have to be dug to about 3 ft in depth. Making a pit could be a lot of work, especially if in the winter when the ground is partially frozen. It is best to take the plant remains to a landfill. If you are a backyard gardener and keep compost the crop debris could be composted.

Vacuum up all small leaves and so on as any plant debris left over could carry insect eggs or overwintering fungal spores. Then, spray the entire growing area with a 10% bleach solution or other disinfectant. This includes the walls, floors, and growing support trays. The growing channels, pond, pots, seedling trays, tomahooks, and plant clips should be soaked in a 10% Clorox solution for several hours.

Disinfectants are oxidizing agents that kill microorganisms. Others besides bleach you may use include "Virkon" (peroxide) and "KleenGrow" (quaternary ammonium). Use safety equipment such as a respirator, disposable gloves, and suit when spraying these compounds as they are irritants to your skin. Never mix bleach with ammonia or acidic solutions as such combinations produce toxic chlorine gases. So, just be careful and there will be no problems! KleenGrow is registered for greenhouse and indoor crop production facilities. Thoroughly wet the surfaces. Use 1.0 oz per gallon of water or 6–8 mL of KleenGrow per liter of water for greenhouse surfaces and equipment. Always read and follow label directions precisely.

INSECT AND DISEASE CONTROL

Exclude insects as much as possible with the use of screens on any intakes that bring in fresh outside air. Sanitation practices including removing any pruning debris and damaged or deformed fruit reduces diseases. After all you would not be happy living in a dirty, cluttered house so keep your plants' living quarters clean also to assist them from becoming sick! Preventing insects from entering will reduce diseases in that many insects suck on plant tissue passing on fungal spores and viruses to your plants. Sucking insects inject viruses into the plants as they rasp or enter the plant tissues with their mouthparts. These insects include aphids, mites, thrips, and whiteflies. They also carry fungal spores on their bodies and transmit them to plants as they suck the juices from the plants creating an ideal point of entry for fungal spore germination.

Prevent diseases by keeping the plants healthy with an active root system. Proper oxygenation, moisture levels, and nutrition all play an important role in healthy, vigorous plants as we discussed earlier. Good hygiene of keeping the growing area clean will reduce the presence of material that may harbor diseases. Be vigilant in recognizing any plant symptoms expressed by the presence of a disease or insect. Early detection and identification of any diseases and/or insects is the key to successful control of these ailments. There are many websites (see Appendix) that give colored pictures of pests and diseases and also recommend control measures. Use these sites for identification and always take photos for future reference.

Take into consideration that the plant symptoms may also be an expression of nutritional or environmental disorders. Plant disorders are discussed in Chapter 10.

As soon as you have determined the cause of the disease or which pest is present, take fast, corrective actions to control them to prevent their spread. Use approved chemical sprays. Seek the use of natural pesticides (bioagents). If these are not sufficiently effective, then apply stronger ones. Do not repeatedly apply the same pesticides in future infestations; vary the type of pesticides to minimize any possible resistance build-up by the pests. An even better approach is to introduce natural predators (beneficial insects) into the crop that will eat or parasitize the pests keeping their numbers limited. These beneficial insects are available through numerous distributors and even hydroponic outlets. Once again there is a lot of information available on the Internet (see Appendix).

COMMON DISEASES

When selecting varieties, as discussed in Chapter 22, you should choose resistant varieties against diseases. The use of resistant varieties will simplify and increase your success. While hydroponic growing greatly reduces the threat of disease in the substrate, it does not prevent diseases in the plant growth above the substrate. Maintaining optimum environmental conditions and controlling pests will help greatly in the prevention of diseases.

Once you think a disease is present start to identify specific areas on the plant that are affected and describe the nature of the symptoms (Figure 25.1). Firstly, identify the area affected: leaves, flowers, fruit, growing tip, stem, crown area, or roots. It may be a combination of these. For example, if the plant wilts during the high light and temperature periods of the day, probably the roots are infected reducing water

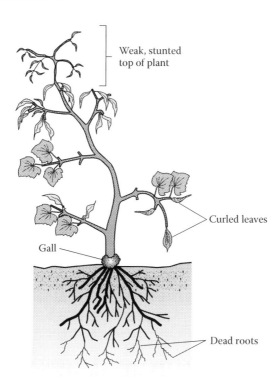

Weak, stunted
top of plant

Curled leaves

Gall

Dead roots

FIGURE 25.1 Disease symptoms on plants. (Drawing courtesy of George Barile, Accurate Art, Inc., Holbrook, New York.)

uptake. Cut some roots to determine whether or not they are turgid and white or soft and slimy. If the latter, you know immediately there is a root problem.

Is the overall plant form stunted or dwarfed? Is the top of the plant very bushy with many small leaves and short internodes? Look for the following symptoms: deformed, wrinkled, rolled, curled, mottled, chlorotic, necrotic leaflets. Look for the presence of spots-concentric or curved, white, powdery, hair-like growth on the leaves (due to some fungi, such as powdery mildew and Botrytis).

Cut the stem on a plant to find out whether or not the vascular tissue is clear and white or brown and soft that would indicate the presence of a disease organism. Any discoloration or softness at the collar (crown) of the plant would indicate a disease. A fruit may be deformed or have spots or lesions pointing to the presence of a disease. After describing and taking photos for future reference of these symptoms go onto websites of the Internet to find photos and descriptions of disease symptoms that may be similar to those of your plants (see Appendix).

TOMATOES

1. *Leaf mold (Cladosporium)*: It starts as a small gray spot on the underside of the leaf and expands into a pale area on the upper surface. Good sanitation,

ventilation, and temperatures preventing high humidity reduce potential infection. Some effective fungicides are available.

2. *Early blight (Alternaria and Septoria)*: Dead spots appear on leaves, attacking older leaves first. Ventilation reduces infection. Remove lower leaves as plants are lowered to increase air circulation thus creating lower relative humidity.

3. *Gray mold (Botrytis)*: These fungal spores enter wounds. That is the reason for cutting leaves with a sharp knife or pruning shears to get a clean surface that will heal quickly. Gray mold appears as a moist rot with a fluffy gray (hairy-like) growth above the infected area. Some fungicides can control the infection at its early stage before the fungus destroys (girdles) the entire plant stem.

4. *Viruses*: Most greenhouse varieties have resistance or tolerance to many viruses. Reduce potential infection by controlling sucking insects that are vectors (aphids, mites, whiteflies).

EGGPLANTS

Since tomatoes, eggplants, and peppers are of the Solanaceous family, they are prone to the same diseases. Tomatoes and eggplants are of the genera *Solanum* and so are very closely related. This family also includes vegetables such as potatoes and some flowers.

1. *Gray mold (Botrytis)*: This is the most prevalent disease of eggplants. Maintain optimum humidity levels through ventilation and temperature. De-leafing lower, yellowing leaves will assist in keeping the humidity near the plant base low. Make a clean break or cut at the base of the leaf petiole (where the leaf joins the stem). *Botrytis* will also affect fruits, stems, and leaves. Cut the fruit during harvesting with a pruning shears or sharp knife to create a rapid healing of the wound. After flowering remove dead flowers that have not set fruit as often *Botrytis* quickly invades these dead tissues.

2. *Stem rot (Sclerotinia)*: This is a fungus that infects the stem of eggplants. Treat it as for *Botrytis*. Practice good sanitation and ventilation.

PEPPERS

Both gray mold and stem rot occur on peppers. Treatments are the same as for eggplants and tomatoes. A number of viruses also may infect peppers. The best step is prevention by elimination of sucking insects.

CUCUMBERS

1. *Powdery mildew*: This is the most common disease on cucumbers. Small white spots appear on the upper leaf surface (Figure 25.2). It spreads rapidly to nearby leaves and plants. Spots enlarge and spread to cover the entire leaf surface as the disease progresses. Proper sanitation and ventilation assist in preventing this disease. Elemental sulfur vaporized by a heater will create a cloud that can penetrate all areas within the crop. This is generally done

FIGURE 25.2 Powdery mildew on cucumber.

overnight. The best remedy for this disease is the selection of resistant or highly tolerant varieties such as Dominica, Logica, and Marillo. Powdery mildew is especially infectious under tropical, humid conditions.

2. *Gray mold (Botrytis)*: The symptoms and controls are similar as for tomatoes.

3. *Gummy stem blight (Didymella bryoniae)*: This disease of flowers, developing fruit, petiole, and base of the main stem is expressed as tan-colored lesions. Good ventilation and optimum relative humidity will discourage infection by this fungus. Some fungicides will arrest the infection.

4. *Cucumber mosaic virus (CMV)*: Some strains of this virus also infect tomatoes. Affected leaves become dwarfed, long, and narrow. There is no cure, only prevention through sanitation and eradication of sucking insects. There are many cucumber cultivars now resistant to this virus. They are indicated by the code CMV after the variety name.

LETTUCE

Lettuce is susceptible to several diseases including *Botrytis*, which is treated as for any of the other crops.

1. *Bacterial soft rot (Erwinia carotovora)*: This is a bacterium. It causes rotting of the internal part of the head as it forms and also at the crown of the plant. Optimum ventilation to maintain optimum humidity levels helps in abating this disease. Sanitation between crops and during production helps to minimize any infection.

2. *Lettuce big vein (Mirafiori lettuce virus)*: Symptoms are enlarged, clear veins of the leaves. Leaves become ruffled and malformed in appearance. Sanitation and resistant varieties are the means of prevention.
3. *Powdery mildew*: This disease produces the same white spots on leaves as in the case of cucumbers. Treatment is the same with vaporized elemental sulfur. Removal of the crop followed by strict sanitation is obligatory. Use resistant varieties.

HERBS

Some herbs are particularly susceptible to Powdery Mildew and *Botrytis*. Basil is affected by bacterial soft rot, especially after harvesting when packaged. *Fusarium* root rot in basil can be avoided through the use of resistant varieties such as "Aroma 2" and "Nufar." Fungal and bacterial leaf spots may occur, especially under high relative humidity or excessive moistening of leaves. Follow control measures as outlined earlier for the other crops.

COMMON PESTS

Most of the pests such as aphids, larvae of caterpillars and moths, mealybugs, two-spotted spider mites, thrips, and whiteflies infest all of the crops. However, some are more aggressive on certain crops than others. They are listed in order of importance under each crop. For details on the identification, life cycles, and control measures refer to websites and books listed in the Appendix such as my book *Hydroponic Food Production*.

Place yellow sticky traps on the overhead wires or support strings about a foot above the top of the plant to catch and monitor the presence of these pests.

TOMATOES

1. *Whiteflies*: This is the most troublesome pest associated with tomatoes (Figure 25.3). You can easily identify this insect by its white wings and body. It is most prevalent on the undersides of leaves and flies quickly when disturbed. There are beneficial insects as well as pesticides available for their control. All of those details you can get at various websites listed in the Appendix.
2. *Aphids*: These pests are almost always found in your backyard garden. They are green, brown, or black in color depending upon the species. Their distinguishing pear-shaped body places them apart from other insects (Figure 25.4). There are winged and wingless forms. One prominent characteristic of their infestation on plants is the presence of "honeydew" excreted from their abdomens causing stickiness of leaves and plant parts as they suck on the plants. This liquid attracts ants, so if you encounter large ant populations around the plants it could be due to the presence of aphids. Often sooty molds (fungi) infect the leaves as a secondary organism, creating a black film on the leaves.

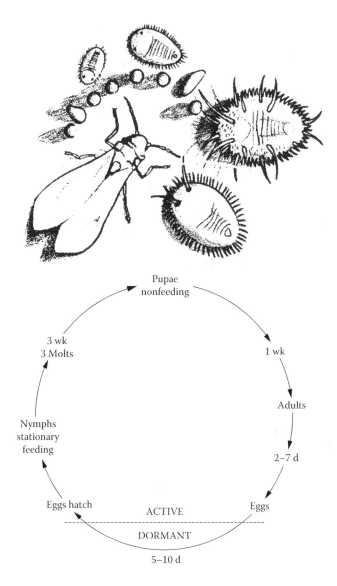

FIGURE 25.3 Whiteflies with life cycle. (Insect drawings courtesy of J.R. Baker, North Carolina Agricultural Extension Service, Raleigh, North Carolina.)

3. *Two-spotted spider mite*: Mites are related to spiders and ticks. They have four pairs of legs compared to insects that have only three pairs of legs. They have two dark-colored spots on their bodies that differentiate them from other mites (Figure 25.5). As they suck on the leaves, small yellow spots form that eventually coalesce to give a bronze appearance to the leaves. They also produce webbing on the leaf surface as infestation increases. If not controlled when numbers are manageable, they will cause complete bleaching and death of the leaves as they suck out all the contents of the cells.

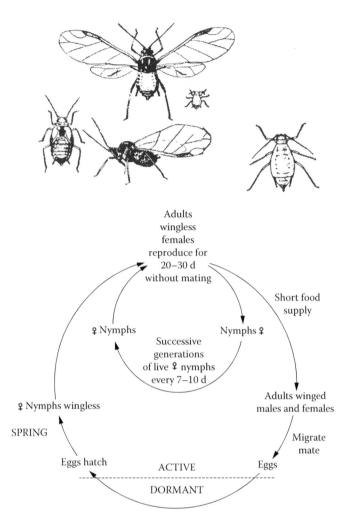

FIGURE 25.4 Aphids with life cycle. (Insect drawings courtesy of J.R. Baker, North Carolina Agricultural Extension Service, Raleigh, North Carolina.)

Several other mites exist that also damage greenhouse crops, carmine mites, and broad mites. These, however, are not as prevalent as the two-spotted mite. They lack the two dark spots and differ in color. The carmine mite is bright red, while the broad mite is translucent and can only be seen with a hand lens. Broad mites cause leaf and fruit deformation.

4. *Thrips*: These insects are especially attracted to the flowers. Their distinct feature is the presence of feathery wings (Figure 25.6). They have rasping mouthparts that scrape the leaf surface and suck the plant sap, causing white, silvery streaks on the leaves. They, like whiteflies and aphids, also carry viruses. Thrips are more attracted to blue sticky traps.

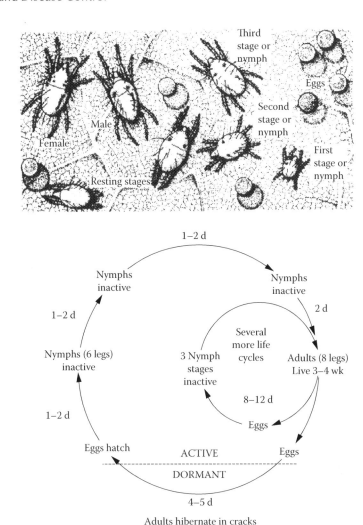

FIGURE 25.5 Two-spotted spider mites with life cycle. (Insect drawings courtesy of J.R. Baker, North Carolina Agricultural Extension Service, Raleigh, North Carolina.)

5. *Leafminers*: Adult leafminers are flies yellow-black in color (Figure 25.7). They deposit eggs in the leaves that show as white swellings. As the larvae hatch, they eat "tunnels" through the leaf between the upper and lower leaf epidermis, creating "mines." As infestation increases, the mines coalesce resulting in large areas of damage that eventually lead to the death of the leaf. The mature larvae drop to the ground (surface of the substrate) where they pupate (go through metamorphosis to adults) within 10 days. The cycle begins all over again.

Reduce infestations by the removal of badly infected leaves and clean up any fallen leaves from the floor. If the substrate is covered with white

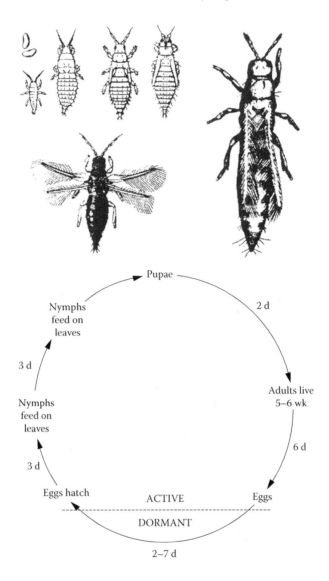

FIGURE 25.6 Thrips with life cycle. (Insect drawings courtesy of J.R. Baker, North Carolina Agricultural Extension Service, Raleigh, North Carolina.)

polyethylene to prevent the larvae from entering as they fall from the leaves, it will minimize the reproduction of the insects. This is particularly helpful if your plants are growing in pots or beds. The use of plastic wrapped slabs will restrict the infestation by breaking the life cycle.

6. *Caterpillars and cutworms*: These are larvae of butterflies and moths, respectively (Figure 25.8). Their presence on crops is indicated by notches in leaves and cut stems and leaf petioles. Cutworms climb up the plants and feed at night, going back to the substrate to hide during the day. Caterpillars feed day

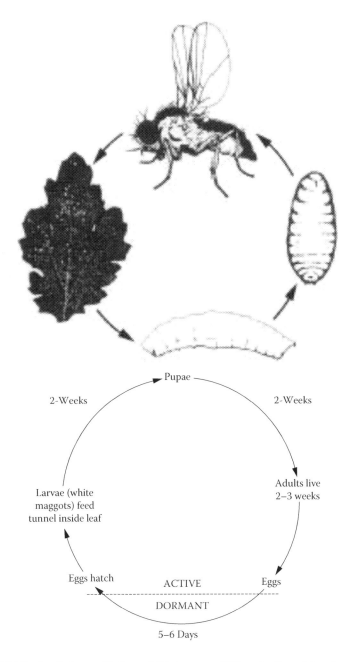

FIGURE 25.7 Leafminers with "tunnels" in leaves and life cycle. (Insect drawings courtesy of J.R. Baker, North Carolina Agricultural Extension Service, Raleigh, North Carolina.)

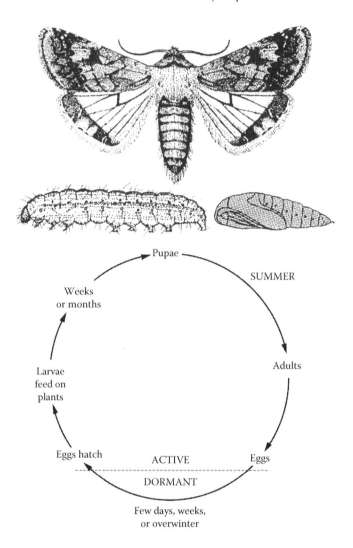

FIGURE 25.8 Caterpillars with life cycle. (Insect drawings courtesy of J.R. Baker, North Carolina Agricultural Extension Service, Raleigh, North Carolina.)

and night. Another tell-tale sign of their presence is the excrements on leaves where they are feeding. Some hornworms can kill an entire plant in one day or night. Look for their signs and pick them off by hand. Also, control them by spraying Dipel or Xentari weekly. Dipel and Xentari are a bacterium (*Bacillus thuringiensis*). This product is very safe and is a biological control agent.

EGGPLANTS

1. *Whiteflies*: These are very common on eggplants. Treatment is as for tomatoes.

2. *Aphids*: The same types of aphids and their control as for tomatoes applies.
3. *Two-spotted spider mite*: Once again their control is the same as for tomatoes.
4. *Mealybugs*: These are notorious on peppers, eggplants, and basil. They have a very characteristic appearance of forming a white wax-like substance covering their bodies (Figure 25.9). It is powder filaments and projections or plates. These filaments protect the insect from contact with many sprays. When using a spray, add a sticker to breakdown the surface tension of the filaments so that the pesticide can contact the insect. There are a few beneficial insects that prey on the pests.
5. *Caterpillars and cutworms*: Look for the same signs as for tomatoes and control them in the same manner.

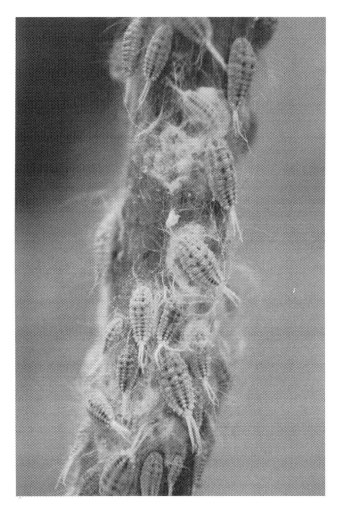

FIGURE 25.9   Mealybugs on pepper stem.

PEPPERS

1. *Mealybugs*: The worst enemy of peppers and the most difficult to control is the mealybug (Figure 25.9). Peppers are also susceptible to whiteflies, aphids, thrips, mites, caterpillars, and cutworms in that order of importance. But these are all easier to control than the mealybugs. Mealybugs cause secretion on the leaves and fruit as they suck the juices from the plants. This sticky substance attracts ants and initiates secondary fungus infection such as sooty mold.

   Control mealybugs as soon as they are sighted at an early stage before they form their protective waxy coating. At this early stage use natural pesticides such as M-Pede, Azatin, and BotaniGard. Infestation progresses very rapidly and will cause leaves to senesce (turn yellow), die, and fall. Remove any heavily infested leaves and bury them.

2. *Broad mites*: Broad mites are translucent and much smaller than the carmine or two-spotted spider mite. The first signs of damage are the curling of young leaves and their becoming brittle. Soon after the initial plant responds, the broad mites will kill the growing point of the plants (Figure 25.10). Once the growing points are damaged to the extent that they dry and break off, the plants are lost as they do not easily form new side shoots. Broad mites also cause scarring of fruit and its deformation. The fruit is not useable at that stage of symptom development.

   Fortunately, they can be controlled by use of Azatin, Neemix, and Abamectin. There are also a number of predatory mites that keep the broad mites in check.

3. *Thrips*: These are more common in peppers than in tomatoes or eggplants. Again monitor them with the sticky cards and control them with natural pesticides and beneficial insects.

4. *Caterpillars and cutworms*: These not only chew holes in the leaves, but also may cut through the stems and eat their way into the fruit at the stem end and then devour the fruit from the inside until it falls from the vines.

CUCUMBERS

1. *Mites*: All three of the mites mentioned earlier love to feed on cucumbers. The two-spotted red spider mite is most common, but carmine and broad mites also attack cucumbers. Broad mites cause the death of the growing point and damage fruit with the appearance of many small white spots that also cause deformation of the fruit. However, malformation is not as severe as with the fruit of peppers. Both beneficial predatory mites and some midges feed on these pests.

   Mites are probably the worst pest of strawberries causing damage on the fruit and leaves with webbing as the infestation progresses.

2. *Whiteflies*: Always check the undersides of leaves as that is generally where infestation begins. Control is as for tomatoes and other crops.

3. *Mealybugs*: Mealybugs probably prefer peppers, eggplants, basil, cucumbers, and tomatoes in that order. At least I have found that in my experience over the years!

FIGURE 25.10 Broad mite damage on peppers.

4. *Aphids*: All types of aphids infest cucumbers.
5. *Thrips*: They are common in cucumbers damaging leaves and fruit.
6. *Caterpillars and cutworms*: These seem to be less of an issue with cucumbers compared to other crops, perhaps due to the rough leaves of the cucumber.

LETTUCE AND ARUGULA

1. *Whiteflies*: These are harsh on lettuce and arugula. Early control is the best approach with beneficial insects and natural chemicals.
2. *Thrips*: Thrips cause severe damage to the leaves scarring the tissue with their rasping mouthparts. The symptoms include the bronzing or silvery streaks on the leaves and stunting of the plants as thrips suck on the young growing point. Thrips prefer flowers or young growing tips of the plants where they hide. When applying pesticides, add a tablespoon of brown sugar per gallon of the tank mixture to bring the thrips out of their hiding places in the tips and in flowers so that a contact pesticide will reach them.

3. *Leafminers*: These can be terrible with lettuce causing loss of leaves and stunting of the plants. Follow the steps outlined earlier for control.
4. *Caterpillars and cutworms*: The best cure here is weekly application of Dipel or Xentari to keep small larvae under control.
5. *Aphids*: All types of aphids will infest lettuce and arugula. Arugula is less troubled by these pests than lettuce, which is more succulent.

HERBS

Most herbs are fairly resistant to many pests. Whiteflies, aphids, mealybugs, and mites prefer basil, thyme, chervil, and chives. Just be vigilant with all of these crops, identify the pests early and treat them as soon as detected. That will make their control much easier with less damage to the plants.

BOK CHOY

The principle pests of bok choy are caterpillars, cutworms, aphids, whiteflies, and leafminers. Spray the bok choy once a week with Dipel or Xentari to prevent the cutworms and caterpillars.

SEEDLINGS

All seedlings when growing in close proximity in cubes are particularly susceptible to fungus gnats especially when a nutrient solution is applied causing algae to grow on the surface of the cubes. Fungus gnats need moisture to deposit their eggs on the surface of the cubes. Larva hatch to grow as a white worm, about ¼″ long, with a black head. The larvae feed on decaying organic matter and algae, but soon attack the seedlings feeding on their roots. Adults have long legs and antennae with one pair of clear wings (Figure 25.11).

Control them by keeping surfaces dry in aisles, and so on of the greenhouse and applying "Vectobac" or "Gnatrol" once a week by soaking the seedlings. Both of these agents are subspecies of the bacterium *Bacillus*. There is a natural nematode (small microscopic eelworm) predator, trade name "Entonem," that is mixed with water and applied as a drench or spray. There is also a predatory mite that feeds on fungus gnat eggs and larvae.

INTEGRATED PEST MANAGEMENT

Integrated pest management (IPM) is the use of beneficial insects, supported with natural pesticides (bioagents) to control pests in the greenhouse. It is a natural balance of pests and beneficial insects keeping the pest population at manageable levels so that very limited plant damage occurs from their presence. It is not to eliminate the pests entirely as that would exclude the beneficial insects from obtaining adequate food to maintain their existence. Beneficial insects may be predators in which they eat their prey (pests) or they may be parasites to the prey by depositing eggs in the prey and as those eggs hatch they actually destroy the pests from the inside using

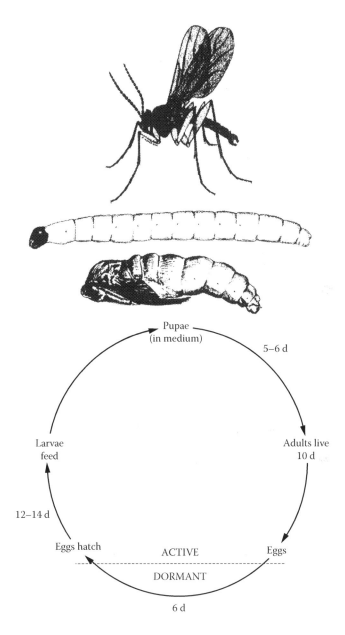

FIGURE 25.11   Fungus gnats with life cycle. (Insect drawings courtesy of J.R. Baker, North Carolina Agricultural Extension Service, Raleigh, North Carolina.)

them as their food source. They are really ferocious and aggressive insects that cause the death of many pests, but always maintaining some balance in retaining sufficient numbers of pests to feed them and their next generations as the cycle repeats.

These beneficial insects are specific to specific pests. Biological supply houses, such as Koppert, rear the beneficial insects and sell them to greenhouse operators. The suppliers of beneficial insects also provide information on monitoring the progress of the beneficials and the control of the pest population. Refer to the Appendix for websites of numerous suppliers of beneficial insects. In most cases, hobby hydroponic growers may use IPM with the assistance of local hydroponic shops or online where these beneficials are available. The key to success is the proper identification of the pest and then obtaining specific beneficial insects to control that pest. Monitoring of the pest population at least several times a week using sticky cards to attract the pests give a representative sample of the numbers and whether or not the population is increasing or remaining stable by the assistance of the beneficials. This monitoring and balancing of both pest and prey is the basis of successful IPM. It is a little more complex method of controlling pests, but is all natural without the use of synthetic pesticides. You may use at times natural pesticide (bioagent) sprays to reduce outbreaks in pest populations to assist the predators in gaining control. However, it is important that these bioagents are not detrimental to the beneficial insects. Information on the compatibility of the bioagents with the beneficials is available on various websites.

SPRAYING

When applying pesticides with a sprayer, choose the correct equipment to match the size of your hydroponic system to make the task easier. In most cases, a backpack sprayer, such as the "Solo" sprayer that has a capacity of 3 gallons, is more than adequate for indoor and backyard operations. It comes with an adjustable nozzle to disperse the spray from a stream to a fine mist. Mix the pesticide with water according to the rate directions exactly as given on the label. Spray during the early morning before temperatures rise and if in a backyard greenhouse before the sunlight

FIGURE 25.12   Burn on a cucumber leaf from a pesticide.

becomes intense, as high temperatures and light can cause burn on the plant leaves. Burn appears as a white clearing of the leaf tissue as shown in Figure 25.12. After spraying you must clean the sprayer thoroughly using soap and water to prevent any residual carry over to your next use, especially when applying numerous different pesticides with the same sprayer. This applies to all pesticides whether natural or synthetic.

SUMMARY

Keep pests and diseases under control in the surrounding area of a backyard greenhouse. If you are also growing in a backyard garden in traditional soil during the spring and summer, this applies to those crops as well. Sanitation in and around the hydroponic system as mentioned earlier is a good practice to reduce pests and diseases. Finally, keep your plants healthy through providing them with optimum nutrition and environment. Healthy plants have thick cuticles and strong tissues that will discourage pests and especially reduce diseases. Avoid rapid vegetative, succulent growth of the plants that causes weak growth and thin tissues that are susceptible to disease infection.

While there are many pests that can damage your crops, their entrance to your crop is more limited indoors than outside in a backyard garden. Nonetheless, once a few enter they will quickly multiply in the ideal environment of your crops. Seek information on many of the helpful websites listed in the Appendix to identify and quickly control the pests. The great advantage of growing indoors or in a greenhouse is that you can introduce natural biological agents that are predators or parasites of the pests and they will live happily in the environment surrounding the crops keeping the pest population in balance. In your backyard garden, this containment of natural agents is difficult, if not impossible, so you must rely more on pesticides. Use natural pesticides (bioagents), whenever appropriate instead of synthetic ones that are persistent in the plants and environment. With hydroponics, the growing area is initially free of all pests and diseases, but as the crop develops, the entrance of such organisms may occur damaging the aerial part of the plants. Watch for them, monitor the area with sticky cards, and act quickly to control them when you encounter them. That will make your growing more productive and enjoyable!

Section VII

Sprouts and Microgreens

26 Sprouts versus Microgreens

SPROUTS

Sprouts and microgreens are very unique crops both in their use and growing. They are used in sandwiches, salads, garnishes, and cooking. Mung bean sprouts are popular in cooking, especially in Chinese and oriental cuisine.

Sprouts of alfalfa, beans, radish, broccoli, and mixtures of alfalfa with onions, garlic, clover, cabbage, fennel, fenugreek, kale, leek, lentil, mustard, cow peas, and green peas are popular nowadays. Many are blended with alfalfa, which makes up 60%–80% of the mix. The seeds of these plants are raw and untreated, and hence the sterilization of their surface is imperative to prevent any possible presence of human transmitted diseases like *Escherichia coli* and *Salmonella*. Purchase the seeds from a reliable source that screens them for the presence of these diseases. For example, Johnny's Selected Seeds are tested to be negative for the presence of these diseases and certified "organic." They also offer a "sprout mix" that contains broccoli, China Rose radish, alfalfa, and Red Russian kale. They describe this mix as "Various shades of green leaves and pink and white stems with a crisp mildly spicy flavor."

If you wish to grow sprouts, it is easiest to purchase a small kitchen unit such as that offered by Johnny's Selected Seeds. They have a "Bioset Germinator" that is specifically designed for germinating seeds. It has a unique siphon action that controls moisture and humidity throughout and is arranged in three layers to separate different crops. It measures about 8″ tall by 6″ in diameter and sells at a cost of $22. They also offer germination kits that are five bags of sprout seeds to fit the germinator.

As mentioned earlier, it is important to sterilize all seeds that will be grown for sprouts and microgreens. Sterilize with about 4000 ppm active chlorine. Use a 10% bleach solution soaking the seeds for about 5–10 min. For example, "Clorox" it has 5.25% active ingredient of sodium hypochlorite. A 10% solution, therefore, will contain 5250 ppm of sodium hypochlorite (1% = 10,000 ppm; so dilution to 10% is 10% × 5.25% × 10,000 ppm = 5250 ppm). This concentration is good for hard seed coats such as those of alfalfa and Mung beans. For microgreens sterilize seeds having softer seed coats, such as amaranth and lettuce, for 4–5 min. Simply watch the bleach solution as you swirl the seeds around in a glass. If the solution starts to become brown, it indicates that the seed coats are starting to break down. At that point the sterilization process is adequate. If you continue longer, there is a risk that the chlorine solution will damage the seed.

Rinse the seeds with raw water several times and place them in the growing containers. Cleanliness is of the highest priority to prevent infection by fungi and bacteria. Use disposable gloves when spreading the seeds.

Both sprouts and microgreens are grown in your home, so temperatures and humidity levels are fine. You may grow them on the kitchen counter or in a bay window. Sprouts do not require any set amount of light, but microgreens should have 14 hours of light of similar intensity to other crops. However, you may use several 30-watt compact fluorescent lights suspended about a foot above the trays to meet their needs. That is much easier than the lighting used with your hydroponic garden. Even though sprouts and microgreens are not thought of as hydroponic, they really are. For that reason, I wish to present their growing methods.

MICROGREENS

Because microgreens are grown under light, they are more nutritious than sprouts. They are generally grown only to their cotyledon stage or slightly longer with the first appearance of true leaves (Figure 26.1). Microgreens grow in the presence of light permitting photosynthesis to occur and the development of chlorophyll (green) and Anthocyanin (purple) pigments. This pigment formation gives the microgreens added nutrients compared to sprouts. Anthocyanin pigment is high in iron content. The other difference between sprouts and microgreens is that with sprouts you eat the shoots and seeds, whereas with microgreens you simply cut off the shoots above the growing substrate leaving the seeds and seed coats behind. This is an added

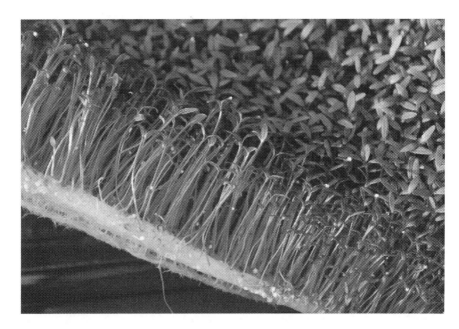

FIGURE 26.1 Amaranth microgreens. (Courtesy of CuisinArt Golf Resort and Spa, Anguilla.)

TABLE 26.1
Potential Microgreen Crops

Amaranth-red	Fennel
Arugula	Kale
Asian greens	Kohlrabi-purple
Basils-Italian, Lemon, Opal	Mint
Beets	Mizuna
Broccoli	Mustard-red
Cabbage	Nasturtium
Celery	Onion
Chives	Parsley
Chervil	Radish-Daikon, red
Cilantro	Spinach
Collards	Sweet pea
Cress-Pepper Cress, Watercress	Swiss chard
Dill	

safety factor in avoiding any illness as the human sicknesses attributed to sprouts are mainly located on the seed coats if not sterilized sufficiently.

Microgreens are very easy to grow and give prolific production. They take a little longer to harvest stage than sprouts, which take 4–5 days. Some radish microgreens will be ready within 5–6 days, but most other crops take up to several weeks. Therefore, when combining different crops of microgreens, you should be aware of the cropping cycles in order to mix those of similar growing periods. In catalogs of microgreen seed providers, the seed varieties are classified into the categories of fast and slow growing. This assists you in the choice of crops to mix within one tray so that they mature at similar times. Seed companies such as Johnny's Selected Seeds also offer several micro mixes as a mild or spicy mixture with varieties that grow well together and give a unique flavor.

Some of the fast-growing varieties include cresses, Chinese cabbage, mustards, Mizuna, radish (Daikon, Hong Vit, and Red Rambo), Pac Choi, Red Russian kale, purple kohlrabi, and Tatsoi. Slow-growing varieties include arugula, beets, sorrel, chard, Komatsuna scallions, Red Garnet Amaranth, basils, orach, and Red Giant mustard. A summary of the most popular microgreens is given in Table 26.1. Try other crops as well. A little experimentation can result in discovering a very appealing crop!

Asian greens include Mitsuba, Pak Choi, Sambuca, Shiso (green and purple), Tangerine, Tatsoi, and Wasabi.

Remember that your choice will depend upon what flavors you prefer and their culinary function. Availability of seeds, such as those of the Asian greens, may be limited. You may have to search websites of Asian seed companies. This is all part of making your hobby hydroponics interesting, unique, and rewarding!

27 Growing Techniques for Microgreens

Often people are somewhat confused as to what are the differences among sprouts, microgreens, and baby greens or mesclun mixes. These classes of products are really differentiated by their size at harvest. In terms of age and size, their categories are as follows: sprouts—youngest, grown in dark, smallest, with the presence of many seed coats; microgreens—somewhat larger in size and older (usually 2″ tall) and grown in full light; baby greens or mesclun mixes—are the oldest and largest (usually 3–4″ tall). Baby greens or mesclun mixes are also cut several times before replanting, whereas, sprouts and microgreens are a one-time harvest. Sprouts are germinated seeds and eaten in entirety (root, seed, and shoot). The edible portions of microgreens are the stems and leaves.

EQUIPMENT AND SUPPLIES

Chapter 26 presented information on sources of seeds and the types of microgreens. Use different seed catalogs and look for a section on microgreens or search websites that have information such as those listed in the Appendix.

Microgreens are easily grown for home use. I have found that the simplest system to grow them is in plastic trays. The following text lists the supplies needed to get started.

PLASTIC TRAYS

These are standard flat trays of 10½″ × 21″ used for propagation. The difference, however, is to obtain these trays without holes. They are available from garden centers and online from greenhouse suppliers such as CropKing (see Appendix). The tray acts as a reservoir to retain moisture for the growing substrate. They may be re-used between crops by sterilizing with a 10% bleach solution.

SUBSTRATE

In my experience with growing microgreens, the best medium is the "Sure To Grow" (STG) mats. These are special capillary mats made for microgreen production. Refer to the Appendix for their website. The only problem with them is that you have to purchase a case of 300 pads for about $200. An alternative is to use several layers of thick paper towels. They are not quite as good as the capillary mats but they will serve the purpose.

SEEDS

As discussed in Chapter 26, these are available from some seed houses such as Johnny's Select Seeds (see Appendix). However, you may use almost any crop that you feel would suit your needs. But one note of caution: purchase only untreated or organic seeds that have no fungicide coating. The use of fungicide on many seeds is to prevent damping off disease of seedlings.

You will note their presence by the seed coat color of yellow, orange, red, green, or other unnatural color of the seed. Some seeds may only be hot-water treated, so they would be okay.

SEED SURFACE STERILANT

Household bleach or Clorox is adequate. As explained in Chapter 26, dilute it to 10% with water. More details of this are presented later.

NUTRIENTS

Purchase a basic vegetable formulation from a hydroponic supplier online or at a shop (see Appendix). Dilute the nutrient solution recommended rate to half strength.

LIGHTS

Make a stand of polyvinyl chloride (PVC) pipe to support two 30-watt compact fluorescent bulbs about 1 ft above the microgreen tray. You could use an aluminum pie plate or other reflective surface just above each light to reflect the light toward the growing tray.

LIGHT TIMER

Use an inexpensive time-clock such as ones used in homes. It should be a 24-hour clock with 1-hour intervals.

MISCELLANEOUS

Measuring cup or graduated cylinder.
A small colander (strainer) to catch seeds as you rinse them during the sterilization procedure.
Disposable gloves to handle the seeds after sterilization.
A teaspoon (tsp) to measure, stir, and spread the seeds on the substrate in the tray.

GROWING PROCEDURES

The most popular seeds used were discussed in Chapter 26. Remember to select seed combinations within any tray that have similar growth rates. That is, avoid mixing slow-growing seeds with fast-growing ones so that your crop develops evenly and at the same pace.

Surface Sterilization

Successful growing depends upon the prevention of diseases from entering the growing tray. For that reason, I do not recommend growing microgreens outside as they will quickly become contaminated with fungi and bacteria that blow in the air with windblown dirt. The best location is on a kitchen counter or in a spare room or basement of your home.

Since you are using raw, untreated seed, the seed surface may contain bacterial or fungal spores that would quickly germinate when the seeds are placed in a moist environment in the growing tray. Surface sterilize with a 10% bleach solution (dilution of one part of bleach to nine parts of water). Make this up in a measuring cup or graduated cylinder (Figure 27.1). Pour the solution into a glass with the seeds and stir with a spoon for 4–5 min. You will have to determine the amount of seeds needed per tray to get the correct density. Be careful not to sow the seed too thick as that will cause crowding of the germinating shoots resulting in weak, elongated stems that are very susceptible to fungal infection. From my experience use 3 level tsps per tray of Amaranth, or lettuce, 2 tsps of Mizuna or Komatsuna, and 7–8 tsps of radish per tray. You will have to experiment with this density to find the best for any specific crop.

Sowing of Seeds

After the sanitation process, rinse the seeds with clean, raw water. Place the seeds in a strainer and rinse. Moisten the paper towels or capillary mat with raw water before scooping the seeds from the strainer with a spoon and spreading them evenly on the

FIGURE 27.1 Surface sterilizing seeds with 10% bleach solution. (Courtesy of CuisinArt Golf Resort and Spa, Anguilla.)

substrate in the tray. Add only sufficient water to fill the grooves in the bottom of the tray until it touches the bottom of the medium. If you fill it too much, the excess water will cause the seeds to float into clumps. You may continue to spread the seeds evenly with your finger as long as you do not touch other things. Add raw water daily for about 3 days until the seeds form small roots (radicals) and shoots (Figure 27.2). Carefully add water at one end of the tray and let it flow by capillary action through the substrate to the other end to avoid floating the seeds. Once the seeds have germinated, their initial roots (radicals) will penetrate the paper towels or mat and secure themselves. Subsequent watering and addition of nutrient solution on a daily basis may be done from above using a watering can with a sprinkler breaker spout, but still take caution in not adding too much water that may cover the seeds. The water level should always be maintained in the tray grooves to the base of the medium.

Within about 3 days when the shoots have grown about ¼–½″ in height, start using a half-strength nutrient solution. Purchase a vegetable formulation concentrate from a hydroponic shop or online and make it up to half the recommended rate per gallon. Store some solution in a dark container of several gallons. The seedlings at this time will have attached to the medium, so they will not drift as you water them. Adding too much solution at a time will inundate them causing lack of oxygen. If you add too much you can tilt the tray to let some drain off. In fact, draining the tray between watering will help to add oxygen to the root mass underneath the medium. Do not use any substrate such as a peatlite or other mix as that will cause excessive moisture levels in the tray and can also introduce substrate getting onto the microgreens during harvesting.

FIGURE 27.2 Radish radical and shoot development into capillary mat after 2 days. (Courtesy of CuisinArt Golf Resort and Spa, Anguilla.)

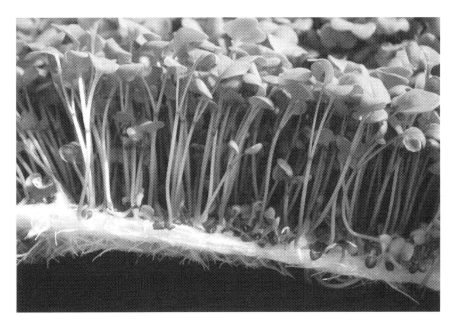

FIGURE 27.3   Radish ready to harvest after 5 days. (Courtesy of CuisinArt Golf Resort and Spa, Anguilla.)

HARVESTING

Microgreens are ready for harvest when they are about 1½–2″ tall at their cotyledon stage (radish) (Figure 27.3) or the initial stage of the first true-leaf formation. Do not allow the first true leaves to expand beyond ¼″ with most varieties. Lettuce true leaves may expand a little more approaching ½″ in length. This stage may vary from 5 days with radish up to 12–14 days with slower-growing varieties such as amaranth. If you want to harvest over a few days begin a little earlier in the cropping cycle. With a scissors cut a section of the substrate with the seedlings and remove them from the tray for immediate use. Cut the seedlings just above the surface of the substrate cutting only the shoots and not the seed coats (Figure 27.4). Collect them in a bowl for use in your favorite salads or other dishes to get a nice added touch to taste and appearance!

Microgreens can be stored for a number of days in the refrigerator if you find that it is easier or that they are growing too fast to harvest daily. They are highly perishable when cut so should be washed and cooled as quickly as possible. Store them in clamshell containers that you could save when purchasing sprouts, Bibb lettuce, or other similarly packaged produce from the supermarket.

TRAY STERILIZATION

After harvesting the microgreens, re-seed more in the same tray(s). However, you must sterilize the tray with a 10% bleach solution for about 20–30 min. Let the tray sit and dry off allowing the chlorine to dissipate before starting the next crop or rinse

FIGURE 27.4 Cutting microgreen (Amaranth) shoots with scissors. (Courtesy of CuisinArt Golf Resort and Spa, Anguilla.)

the tray with raw water. Repeat this process again as instructed earlier. Once again try different combinations of crops to find which grow best and which flavors you like. This is part of the fun of growing them! Always follow the sterilization steps precisely to avoid any contamination and you will get fantastic crops every time!

SUMMARY

You can get ideas on combinations and types of microgreens from numerous websites (see Appendix). One particularly informative website is www.freshorigins.com.

Growing microgreens, like most hydroponic crops, exemplifies plant growth and development combined with plant nutrition. This is a great theme for science projects for young, grade students. A display of growing microgreens will show the germination of plants from seeds, basic requirements of water, oxygen, and nutrients for plant growth. If the microgreens are grown a little longer, up to several weeks, nutritional studies with various nutrient levels or the deliberate non-use of the plant essential elements will be expressed in the growth differences among the trials. All of these aspects of plant growth exemplify to students the application of biology and chemistry and will attract their interest in the sciences as they see it applied. Encourage your children to get involved with you in hydroponic growing! They will better understand and appreciate that fresh produce does not just appear on supermarket shelves, and it will therefore help them comprehend the science, effort, and love for growing plants!

Final Advice on How to Get Started

Do not hesitate, anyone can grow hydroponically. In fact, if you have any indoor potted houseplants like mums, foliage ornamental plants, African violets, or any other flowering potted plants; they are all in a peatlite hydroponic substrate. That is the reason you must water them with a plant food as there are no residual nutrients in the medium apart from those given to them in the greenhouse while growing. So, with these plants you are already growing hydroponically!

Start by visiting some websites (see Appendix) such as my website: www.howardresh.com. These websites will provide lots of information and photos. Also, go to websites of hydroponic suppliers to become familiar with products, supplies, and small units (see Appendix). These websites have lots of ideas and give costs of their products. There are also many books and articles available on the Internet (see Appendix).

Once you have explored these websites decide on what size of system you wish to start with. You may begin with several pots of simple systems such as were shown in Chapter 12. Work your way up to larger units as you gain experience and confidence in various growing methods. Try different systems including wick, floating water culture, and drip irrigation of various substrates. Of course, the size of larger indoor units will be a factor of the room you have available in your home. Become familiar with environmental components as lighting, carbon dioxide enrichment, temperature regulation, and ventilation. You will employ these components as you expand a specific area isolated for hydroponic growing. A basement is an ideal environment in that it is cooler and can be secluded from the rest of the house allowing isolation of the environment in order to regulate these factors important in growing.

Begin with open systems collecting the leachate for application later to your outdoor gardening or its disposal. This will greatly simplify the nutrition of the plants initially allowing greater success until you become more experienced with the management of the nutrient solution. After several crops, modify the system to recirculation of the nutrient solution conserving water and nutrients. In this way, you will have developed the skills of maintaining and adjusting EC and pH levels with the growth of the crops.

Hydroponics is fascinating in that there are many systems of growing. Start with the simpler systems and progress to those more detailed. Broaden your knowledge with the expansion of the hydroponic facility and types of crops. Start with the easier crops like lettuce, arugula, basil, and herbs and continue later with the vine crops of tomatoes, peppers, eggplants, and European cucumbers that require specific training techniques. I am not suggesting that this is the only way to get started successfully, but recommending the route to slowly gain experience and confidence so that you can build on these skills to prepare to grow any crop with hydroponics! It is not rocket science, it is applying your growing skills to become successful with these crops under a controlled environment that will give you maximum yields. The care of your crops in controlling many factors at optimum levels, not possible outside, will reward you with high production and quality of home-grown vegetables. These results are achieved with less labor indoors than in a traditional backyard soil garden.

If you have the space outside the next step you can take is to set up a backyard hydroponic greenhouse for year-round growing. That is another level of hydroponic growing that takes you into the regulation of the greenhouse environment through the use of your preferred structure and components to provide an optimum environment for plant growth. Greenhouse culture is a very rewarding endeavor in self-satisfaction of growing and is an escape from everyday work-related stresses. The greenhouse provides fresh, nutritious crops year round. There is nothing like arriving home from work in the dark of winter months and transforming your senses into a tropical paradise of the jungle of crops growing under bright lights! Try it!

REFERENCES

Many books on hydroponics are available from hydroponic suppliers online. Look for them at hydroponic stores, garden centers and on the Internet such as www.amazon.com and www.barnesandnoble.com.

Bridwell, R. 1990. *Hydroponic Gardening*, Rev. Ed. Woodbridge Press, Santa Barbara, CA.
Cooper, A. 1979. *The ABC of NFT*, Grower Books, London.
Dalton, L. and R. Smith. 1984. *Hydroponic Gardening*, Cobb Horwood Publications, Auckland.
Douglas, J.S. 1984. *Beginner's Guide to Hydroponics*, New Ed., Pelham Books, London.
Douglas, J.S. 1985. *Advanced Guide to Hydroponics*, New Ed., Pelham Books, London.
Harris, D. 1986. *Hydroponics: The Complete Guide to Gardening without Soil: A Practical Handbook for Beginners, Hobbyists and Commercial Growers*, New Holland Publishers, London.
Jones, J.B. Jr. 1997. *Plant Nutrition Manual*, CRC Press, Boca Raton, FL.
Jones, J.B. Jr. 1999. *Soil and Plant Analysis*, CRC Press, Boca Raton, FL.
Jones, J.B. Jr. 1999. *Soil Analysis Handbook of Reference Methods*, CRC Press, Boca Raton, FL.
Jones, J.B. Jr. 2001. *Laboratory Guide for Conducting Soil Tests and Plant Analysis*, CRC Press, Boca Raton, FL.
Jones, J.B. Jr. 2004. *Hydroponics: A Practical Guide for the Soilless Grower*, 2nd Ed., CRC Press, Boca Raton, FL.
Jones, J.B. Jr. 2007. *Tomato Plant Culture: In the Field, Greenhouse and Home Garden*, 2nd Ed., CRC Press, Boca Raton, FL.
Jones, L., P. Beardsley, and C. Beardsley. 1990. *Home Hydroponics...and How to Do It!*, Rev. Ed., Crown Publishers, New York, NY.
Kenyon, S. 1992. *Hydroponics for the Home Gardener*, Rev. Ed., Key Porter Books Ltd., Toronto, Canada.
Marlow, D.H. 1993. *Greenhouse Crops in North America: A Practical Guide to Stonewool Culture*, Greenhouse Gardening, Milton, Ontario, Canada.
Mason, J. 1990. *Commercial Hydroponics*, Kangaroo Press, Kenthurst, NSW, Australia.
Mason, J. 2000. *Commercial Hydroponics: How to Grow 86 Different Plants in Hydroponics*, Simon & Schuster, Australia.

Morgan, L. 1999. *Hydroponic Lettuce Production*, Casper Publications, NSW, Australia.

Morgan, L. 2005. *Fresh Culinary Herb Production*, John Wiley & Sons, Australia.

Morgan, L. 2006. *Hydroponic Strawberry Production: A Technical Guide to the Hydroponic Production of Strawberries,* Suntec (NZ) Ltd. Publications, Tokomaru, New Zealand.

Morgan, L. 2008. *Hydroponic Tomato Crop Production*, Suntec (NZ) Ltd. Publications, Tokomaru, New Zealand.

Morgan, L. and S. Lennard. 2000. *Hydroponic Capsicum Production*, Casper Publications, NSW, Australia.

Muckle, M.E. 1982. *Basic Hydroponics*, Growers Press, Princeton, BC.

Muckle, M.E. 1998. *Hydroponic Nutrients—Easy Ways to Make Your Own*, Rev. Ed., Growers Books, Princeton, BC.

Nelson, P.V. 1998. *Greenhouse Operation and Management*, 5th Ed., Prentice-Hall, Upper Saddle River, NJ.

Nicholls, R.E. 1990. *Beginning Hydroponics: Soilless Gardening: A Beginner's Guide to Growing Vegetables, House Plants, Flowers, and Herbs without Soil*, Running Press, Philadelphia, PA.

Pranis, E. and J. Hendry. 1995. *Exploring Classroom Hydroponics*, National Gardening Association, Inc., South Burlington, VT.

Resh, H.M. 1990. *Hydroponic Home Food Gardens*, Woodbridge Press, Santa Barbara, CA.

Resh, H.M. 1993. *Hydroponic Tomatoes for the Home Gardener*, Woodbridge Press, Santa Barbara, CA.

Resh, H.M. 1998. *Hydroponics: Questions & Answers—For Successful Growing*, CRC Press, Boca Raton, FL.

Resh, H.M. 2012. *Hydroponic Food Production*, 7th Ed., CRC Press, Boca Raton, FL.

Resh, H.M. 2013. *Hobby Hydroponics*, 2nd Ed., CRC Press, Boca Raton, FL.

Roberto, K. 2003. *How-to Hydroponics*, 4th Ed., Futuregarden Press, Farmingdale, NY.

Romer, J. 2000. *Hydroponic Crop Production*, Simon & Shuster, Australia.

Ross, J. 1998. *Hydroponic Tomato Production: A Practical Guide to Growing Tomatoes in Containers*, Casper Publications, Narabeen, Australia.

Schwarz, M. 1968. *Guide to Commercial Hydroponics*, Israel University Press, Jerusalem, Israel.

Schwarz, M. 1995. *Soilless Culture Management. Advanced Series in Agriculture*, Vol. 24, Springer-Verlag, Berlin, NY.

Smith, D.L. 1987. *Rockwool in Horticulture*, Grower Books, London.

Straver, W.A. 1993. *Growing European Seedless Cucumbers*, Ministry of Agriculture and Food, Parliament Buildings, Toronto, Ontario, Canada, Factsheet, Order No. 83-006, AGDEX 292/21.

Sundstrom, A.C. 1989. *Simple Hydroponics—For Australian and New Zealand Gardeners*, 3rd Ed., Viking O'Neil, South Yarra, VIC, Australia.

Taylor, J.D. 1983. *Grow More Nutritious Vegetables without Soil: New Organic Method of Hydroponics*, Parkside Press, Santa Ana, CA.

Van Patten, G.F. 1990. *Gardening: The Rockwool Book*, Van Patten Publishing, Portland, OR.

Van Patten, G.F. 2004. *Hydroponic Basics*, Van Patten Publishing, Portland, OR.

Van Patten, G.F. 2007. *Gardening Indoors with Soil and Hydroponics*, 5th Ed., Van Patten Publishing, Portland, OR.

Appendix: Sources of Supplies and Information

WEBSITES

HOBBY HYDROPONIC GROWING

There are many websites with useful information on hydroponic culture of various crops. Use search engines such as Google for these websites under "Simple or Hobby Hydroponics." Here are a few to get started.

http://hydroponics123123.multiply.com/journal/item/5
http://hydroponics123123.multiply.com/journal/item/6
http://hydroponics123123.multiply.com/journal/item/9
http://hydroponics123123.multiply.com/journal/item/10
http://www.hydro-unlimited.com/index.php?p=2_1
http://www.simplyhydro.com/freesys.htm
http://www.hydroponics-simplified.com/hydroponic-setups.html
http://www.homehydrosystems.com/hydroponic-systems/systems.html
http://www.hydroponicsequipment.co
http://www.aces.edu/pubs/docs/A/ANR-1151/index2.tmpl
http://www.squidoo.com/backyardhobbygreenhouses

HOBBY HYDROPONIC SUPPLIES

There are many suppliers of hydroponic nutrients, equipment and accessories. Here are a few available through the Internet.

Company	Website
Advanced Nutrients	www.advancednutrients.com
AeroGarden	www.aerogarden.com
American Hydroponics	www.amhydro.com
Apache Tech Inc.	www.apachetechinc.com
Aquatic Eco-Systems, Inc.	www.AquaticEco.com
Autogrow Systems Ltd.	www.autogrow.com
AutoPot	www.autopot.co.uk
Better Grow Hydroponics	www.bghydro.com
Bluelab Corporation Limited	www.getbluelab.com
Botanicare (American Agritech)	www.botanicare.com
	www.americanagritech.com
CO2Boost	www.co2boost.com

CropKing, Inc.	www.cropking.com
Current Culture H2O	www.cch2o.com
General Hydroponics	www.generalhydroponics.com
Greentrees Hydroponics	www.hydroponics.net
Grodan	www.grodan101.com
Growco Garden Supply	www.4hydroponics.com
Hanna Instruments	www.hannainst.com
Homegrown Hydroponics	www.homegrownhydro.com
Hort Americas	www.hortamericas.com
Horti-Control	www.horticontrol.com
HTG Supply	www.htgsupply.com
Hydrodynamics International, Inc.	www.hydrodynamicsintl.com
Hydrofarm Horticultural Products	www.hydrofarm.com
Hydrofogger	www.hydrofogger.com
Hydrologic Purifications Systems	www.hydrologicsystems.com
Hydrotek	www.hydrotek.ca
Milwaukee Instruments	www.milwaukeeinstruments.com
Myron L Company	www.myronl.com
Nickel City Wholesale Garden Supply	www.ncwgs.com
North American Hydroponics	www.wearehydro.com
P.L. Light Systems	www.pllight.com
Pulse Instruments	www.pulseinstrument.com
Simply Hydroponics and Organics	www.simplyhydro.com
Solis Tek Digital Ballasts	www.solis-tek.com
Sunburst Hydroponics	www.4hydro.com
Sunleaves Garden Products	www.sunleaves.com
Sunlight Supply, Inc.	www.sunlightsupply.com
Sure To Grow	www.suretogrow.com
Verti-Gro, Inc.	www.vertigro.com

BACKYARD GREENHOUSES AND COMPONENTS

The following companies and websites offer structures, components, and information on backyard greenhouses.

Company	Website
ACF Greenhouses	www.littlegreenhouse.com
Acme Engineering and	www.acmeag.com
Manufacturing Corp.	www.acmehort.com/products.asp
Backyard Greenhouses	www.backyardgreenhouses.com
BC Greenhouse Builders Ltd.	www.bcgreenhouses.com
Gothic Arch Greenhouses	www.gothicarchgreenhouses.com
Grainger, Inc.	www.grainger.com
Intermatic, Inc.	www.intermatic.com
International Greenhouse Company	www.igcusa.com
	www.greenhousemegastore.com

J. Orbesen Teknik ApS.	www.greenhouse-vent-opener.com
Mardenkro	www.mardenkro.com
Modine Manufacturing Co.	www.modine.com
Sherry's Greenhouse	www.sherrysgreenhouse.com/oldsite/GHcontents.html
The Greenhouse Catalog	www.greenhousecatalog.com
Thermoforce Ltd.	www.thermoforce.co.uk/autovents.htm

HYDROPONIC INFORMATION—GOVERNMENT AND UNIVERSITIES—WEBSITES

www.aceis.agr.ca
www.ag.arizona.edu/hydroponictomatoes
www.colostate.edu/Dept/CoopExt/Adams/gh/pdf/dbghobby.pdf
www.cals.cornell.edu/dept/flori/cea/programs.html
www.greenhouseinfo.com
www.usda.gov
www.ontariogreenhouse.com
www.agf.gov.bc.ca/croplive/cropprot/prodguide.html
www.bcgreenhouse.ca/publications.htm
www.ces.ncsu.edu
www.ipm.ucdavis.edu
www.msucares.com/pubs/

PESTS AND DISEASES—WEBSITES OF IDENTIFICATION AND CONTROL—IPM

www.koppert.nl
www.anbp.org
www.intertechserv.com
www.biobest.be
www.mycotech.com
www.bioworksbiocontrol.com
www.biocontrol.entomology.cornell.edu/
www.ipm.ucdavis.edu/
www.res2.agr.ca/harrow/bkindex.htm
www.cips.msu.edu/biocontrol/research.htm
www.nysipm.cornell.edu/publications/greymold.html
http://207.5.71.37/biobest/en/teelten/tomaat.htm
www.naturescontrol.com/controls.html

SOURCES OF SEEDS

Seed Houses

Website

De Ruiter Seeds Inc.	www.deruiterusa.com
(Monsanto Vegetable Seeds)	www.monsanto.com

Johnny's Selected Seeds	www.johnnyseeds.com
Ornamental Edibles	www.ornamentaledibles.com
Paramount Seeds Inc.	www.paramountseeds.com
Richters Herbs	www.richters.com
Stokes Seeds Ltd.	www.stokeseeds.com
Rijk Zwaan USA	www.rijkzwaan.nl
Tainong Seeds, Inc.	www.tainongseeds.com

HYDROPONIC ORGANIZATIONS

Hydroponic societies promote new technology and products. They hold annual meetings or conferences providing up-to-date information from experts within the field of hydroponics. The meetings are very informative introducing new techniques and products. In addition, you will meet people who have been inspired by hydroponics. Most conferences have a Hydroponic Suppliers' Trade show displaying products offered by companies.

ADDRESSES AND WEBSITES/E-MAILS FOR HYDROPONIC SOCIETIES

Address	Website/E-mail
Hydroponic Society of America (HSA) P.O. Box 1183 El Cerrito, CA 94530	www.lisarein.com/hydroponics
Asociacion Hidroponica Mexicana A.C.	www.hidroponia.org.mx
Australian Hydroponic and Greenhouse Association (AHGA) (renamed: Protected Cropping Australia) (PCA) Narrabeen, NSW, 2101, Australia	www.protectedcroppingaustralia.com
Centro de Investigacion de Hidroponia y Nutricion Mineral Univ. Nacional Agraria La Molina Av. La Universidad s/n La Molina Lima 12, Peru	www.lamolina.edu.pe/hidroponia
Centro Nacional de Jardineria Corazon Verde in Costa Rica	www.corazonverdecr.com
Encontro Brasileiro de Hidroponia	www.encontrohidroponia.com.br
Singapore Society for Soilless Culture (SSSC) #13-75, 461 Crawford Lane, Singapore 190461	

INTERNET CHAT CLUBS

There are hydroponic forums where you may sign up to be part of discussion groups online. You can seek advice from other growers and hydroponic experts. Such forums disperse knowledge of new products.

Of course, you can share your experiences with others on websites like YouTube, Facebook and Twitter, and so on.

WEBSITES/E-MAILS OF HYDROPONIC FORUMS

hydroforum@fesersoft.com
hydrolist@hydropoinics.org
http://forums.gardenweb.com/forums/hydro/
www.hydrohangout.com
www.hobbyhydro.com

HYDROPONIC MAGAZINES

While these magazines present both poplar and technical articles, they are written for ease of understanding. The magazines have extensive advertising by manufacturers and suppliers of hydroponic products to keep you informed of new products.

ADDRESSES AND WEBSITES OF HYDROPONIC MAGAZINES

Practical Hydroponics and Greenhouses   www.hydroponics.com.au
Casper Publications Pty Ltd.
P.O. Box 225
Narrabeen, 2101
Australia

The Growing Edge Magazine   www.growingedge.com
P.O. Box 1027
Corvallis, OR 97339

Maximum Yield Gardening   www.maximumyield.com
2339 Delinea Place
Nanaimo, BC
Canada V9T 5L9

National Gardening Association   http://assoc.garden.org
1100 Dorset Street
South Burlington, VT 05403

The Indoor Gardener Magazine   www.theindoorgardener.ca
Green Publications
P.O. Box 52046
Laval, Quebec
Canada H7P 5S1

deRiego   www.revistaderiego.com.mx
Revista deRiego
Apdo. Postal 86-200
Mexico, D.F. C.P. 14391

Urban Ag News   www.urbanagnews.com
Urban Garden Magazine   http://urbangardenmagazine.com

Index

A

Abortion of fruit, 58
Aeroponic seed potatoes, 140
A-frame NFT system
    assembly, 114
    materials, 111–114
Air circulation, 144–145
Ammonium molybdate, 43
Aphids
    arugula, 294
    eggplants, 291
    lettuce, 294
    tomatoes, 284
Arugula
    cultivation of, 257–258
    hydroponic culture, 225
    seedlings, 245–246
AutoPot system, 125–126

B

Backyard greenhouses
    annual production, 190–191
    claddings, 167–169
    components, 179–180
    construction, 170–179
    control panel, 191–193
    cooling costs, 190
    coverings, 167–169
    crop layout plan, 191
    heating, 180–184
    irrigation controllers, 219–221
    lighting components, 216–219
    location, type, and shape, 163–165
    nature of structure, 165–166
    projected annual revenues, 193–194
    site preparation, 169–170
    sizes and prices, 166–167
    time-clocks, 219–221
    total estimated heating, 190
    ventilation, 184–190
Bacterial soft rot, 283
Basils
    cultivation of, 258–259
    hydroponic culture, 228–230
    seedlings, 248–249
Bato buckets, 197–198
Beit alpha (BA) cucumbers
    cultivation of, 272–277

    hydroponic culture, 240
    seedlings, 250–253
Bird's eye gravel, 64–65
Blossom end rot, 55–56
Blotchy ripening, 56
Bok choy
    cultivation of, 259
    hydroponic culture, 226–227
    pest control, 294
    seedlings, 246
Boric acid, 43
Boron
    causes and remedies in leaves, 54
    in plants, 35
Broad mites, 292

C

Calcium
    causes and remedies in leaves, 53
    in plants, 35
Calcium nitrate, 42
Carbon dioxide enrichment
    environmental control components,
        145–146
    greenhouse environment, 156–157
Caterpillars
    arugula, 294
    eggplants, 291
    lettuce, 294
    peppers, 292
    tomatoes, 288–290
Catfacing, 57–58
"Chinese Leeks/Garlic" chives, 231
Chlorine, in plants, 36
Cilantro seeds, 231
Claddings, 167–169
Clamping eggplant, 265
CMV, *see* Cucumber mosaic virus
Coco coir substrates, 67
Coco coir system
    large indoor hydroponic units, 125
    small indoor hydroponic units, 91
Commercial hydroponics, 5
Control panel, 191–193
Copper
    causes and remedies in leaves, 54
    in plants, 36
Copper sulfate, 43
Coverings, 167–169

Printed in the United States
by Baker & Taylor Publisher Services